Emerging Technologies for Smart Cities

Oluwayemi-Oniya Aderibigbe • Trynos Gumbo
Samson Olawale Fadare

Emerging Technologies for Smart Cities

Sustainable Transport Planning in the Global North and Global South

Springer

Oluwayemi-Oniya Aderibigbe
Department of Urban and Regional Planning, Faculty of Engineering and Built Environment
University of Johannesburg
Johannesburg, South Africa

Trynos Gumbo
Department of Urban and Regional Planning, Faculty of Engineering and Built Environment
University of Johannesburg
Johannesburg, South Africa

Samson Olawale Fadare
Obafemi Awolowo University
Ile-Ife, Nigeria

ISBN 978-3-031-66942-2 ISBN 978-3-031-66943-9 (eBook)
https://doi.org/10.1007/978-3-031-66943-9

This Springer imprint is published by the registered company Springer Nature Switzerland AG
The registered company address is: Gewerbestrasse 11, 6330 Cham, Switzerland

Foreword

This book—*Emerging Technologies for Smart Cities: Sustainable Transport Planning in the Global North and Global South* —explores lasting solutions to contemporary urban transportation problems in cities in both higher and lower income countries. Its aim is to identify effective and feasible practices, drawing from existing and emerging technologies, capable of producing smart and sustainable transport systems.

The book is the product of extensive research of possible practices, and their empirical application. These empirical applications draw from the many world cities that are currently grappling with climate change, and with the adoption of measures aimed at achieving smart and sustainable development. The book addresses issues that transcend transport system operations, economic development, climate change, liveability, emerging technologies, and citizen participation.

The book starts with an unpacking of the transport contexts of higher income cities (paying particular attention to cities in Germany, the United States, and Singapore), and then of lower income cities (paying particular attention to cities in Nigeria, South Africa, and India). From this rich diversity of contexts, it moves on to explore conceptualisations of sustainable and smart transport systems. The latter part of the book explores the diffusion, and potential, of existing and emerging technologies, and their applications in transportation and traffic management, and urban transport planning. The authors highlight the differences in context and policy in countries with greater and lesser resources, and how these manifest in disparities in infrastructure quality, public transportation services, and the adoption of advanced technology.

The authors provide an apposite resource for the consultants, government officials, and academics active in the array of disciplines that plan, design, manage, maintain, and regulate built environments. Policymakers, researchers, and university students alike will find its contents useful.

University of Cape Town
Cape Town, South Africa

Roger Behrens

Preface

Technology today is an inevitable necessity for organisation productivity, development of industries, economic growth, increment in a human development capacity, and livelihood improvement. These modern technologies such as Drones, Electric Vehicles, Geographic Information System, Autonomous Vehicles, and Robotics have been very useful in urban planning and efficient management of transportation in the built environment. It has also assisted professionals in the built environment in optimising their operations to boost the delivery of cost-effective, timely, and quality projects aimed at sustainable planning. More importantly, the adoption of these technologies has helped major cities in the global north and south of the world in transforming their society to a smart city.

Smart technologies are increasingly used in different businesses and industries. A wide range of these technologies and digital innovations are applicable to smart cities. In addition, the built environment also explores the benefits associated with these emerging technologies to enhance safety and efficient planning and management of infrastructures. At some point in the city planning processes, the adoption of these technologies and other measures aimed at sustainable planning, decrease emissions, and improve efficiency which can further be harnessed.

However, the adoption of these technologies has not been fully integrated in the transportation planning of most countries in the developing world due to challenges on the part of the citizens and government. To fully harness the benefits associated with these technologies, there is a need for infrastructural development, citizen engagement, and government intervention in the areas of resources allocation, funding, and implementation of appropriate policies.

Johannesburg, South Africa — Oluwayemi-Oniya Aderibigbe
Johannesburg, South Africa — Trynos Gumbo
Ile-Ife, Nigeria — Samson Olawale Fadare

Contents

About the Authors

Oluwayemi-Oniya Aderibigbe is a Professional Urban Transportation Planner and a Lecturer in the Department of Logistics and Transport Technology, Federal University of Technology Akure, Nigeria. She is also a Postdoctoral Fellow in the Department of Urban and Regional Planning within the Faculty of Engineering and the Built Environment in the University of Johannesburg (UJ). She holds a PhD in Transport Management Technology, Federal University of Technology, Akure, Nigeria, as well as a master's and bachelor's degree in Urban and Regional Planning, Obafemi Awolowo University, Ile-Ife, Nigeria. Her research interests include urban transportation planning, travel behaviour studies, sustainable and smart cities, climate change, emerging technologies in urban planning, and green economy. Her research interests leverage a combination of experimental and empirical approaches using both qualitative and quantitative data to study complex problems in the field of urban planning. She has published in reputable chapters, peer-reviewed journals, and peer-reviewed conference proceedings.

Trynos Gumbo is a C2 NRF rated researcher and a SACPLAN Registered Professional Planner, who is currently working as a Full Professor and leader of the Sustainable and Smart Cities and Regions (SSCR) group within the Department of Urban and Regional Planning within the Faculty of Engineering and the Built Environment at the University of Johannesburg (UJ), South Africa. He holds a PhD from Stellenbosch University, South Africa, as well as master's and honours degrees from the University of Zimbabwe (UZ), Zimbabwe. He has previously worked in the Africa Institute of South Africa of the Human Sciences Research Council as a research specialist and Acting Head for the sustainable development programme. Prof. Gumbo has also worked as an international instructor in the urban management master's programme within the Ethiopian Civil Service University College (ECSUC) in Addis Ababa in Ethiopia. Before, Prof. Gumbo had worked as lecturer and Head of Department at the National University of Science and Technology (NUST) in Zimbabwe. He has attended and presented at several national and international conferences and has published widely in a variety of research areas that include informality, housing, urban planning, development, and management. His research interests include urban transportation planning and management, sustainable and smart cities development, housing and economic informality, green economy and renewable energy generation from waste, and innovative building technologies and materials.

Samson Olawale Fadare is a professional planner and retired professor of Urban and Regional Planning in the Department of Urban and Regional Planning, Obafemi Awolowo University, Ile-Ife. He holds a PhD in Urban and Regional Planning, University of Sheffield, UK, as well as master's and bachelor's degree from the Eastern Washington University, USA. During his service at the Obafemi Awolowo University, Ile-Ife, he served as the Head, Department of Urban and Regional Planning, Ile-Ife, and was also the Dean, Faculty of Environmental Design and Management. He was appointed as the chairman committee of Deans for two (2) years and he was a member of the governing council of the Obafemi Awolowo University Ile-Ife, Nigeria. Prof. Fadare was also a Fulbright Scholar in the United States. He has attended and presented at several national and international conferences and has published widely in a variety of peer-reviewed journals. His research interests include urban transportation planning, human development, and urban sustainability.

Part I
Introduction and Setting the Scene

Chapter 1
Emerging Technologies and Concept of Transportation and Development

Oluwayemi-Oniya Aderibigbe and Trynos Gumbo

1.1 Background to the Study

The progress of a country is heavily reliant on its development, which is instrumental in providing its population with a high standard of living. Nevertheless, in order for a nation to embark on a path of development, it must fulfill specific requirements, such as attaining socio-political and economic stability (Lawal & Abe, 2011). A critical element contributing to economic prospects is the existence of a well-established transportation infrastructure capable of satisfying mobility demands and ensuring accessibility to markets and resources (Rodrigue et al., 2006). The development of transportation within a nation relies on the establishment of a comprehensive, streamlined, reliable, affordable, inclusive, and secure transport system. This system plays a crucial role in driving economic, social, and political advancements. It is note-worthy that transportation is a demand that arises from the activities of other productive sectors, rather than being productive in itself. The prosperity of a nation is closely linked to its ability to facilitate trade and distribution. However, it is important to acknowledge that transportation can also give rise to negative outcomes such as traffic congestion, accidents, and disparities in mobility (Rodrigue et al., 2006). According to Adeniran and Yusuf (2016), the significance of transportation in the development of any nation can be summarized as follows:

- Provision of infrastructure that enhances the physical appearance of cities and improves accessibility and mobility.
- Improvement in welfare.

Considering the above, transportation can be seen as the engine driving the growth and development of cities.

A city can therefore be defined as a human settlement that includes a range of infrastructures, such as sewage and water treatment facilities, police and fire

O.-O. Aderibigbe et al., *Emerging Technologies for Smart Cities*,
https://doi.org/10.1007/978-3-031-66943-9_1

departments, utilities, educational institutions, libraries, commercial establishments, and housing. On the other hand, a smart city represents a modern urban concept that emphasizes the involvement of citizens and the incorporation of technology into the city's infrastructure. The main goal of developing a smart city is to enhance the quality of life by utilizing technology to increase service efficiency and meet the needs of its inhabitants (Kim et al., 2018). Cities are no longer just engines of innovation and economic growth; they have also become centers of numerous environmental challenges for our planet. Both developing and developed countries face issues of overburdened and overcrowded cities that give rise to significant societal problems. Cities and counties face a range of challenges, including unemployment, stunted economic growth, poverty, outdated infrastructure, traffic jams, high crime levels, insufficient clean water supply, environmental risks, and sluggish bureaucratic processes. Additionally, cities encounter substantial cybersecurity threats, including cyber-terrorism that could severely impact infrastructure and residents' safety. It is crucial to develop intelligent solutions that enhance livability, promote economic development, and reduce resource consumption.

Urbanization, resulting from the rapid migration of people to urban areas since the industrial revolution, has led to the rapid expansion of cities. This process describes the increasing population in urban centers, with hundreds of thousands of new homes being built every day. Globally, millions of individuals move to cities each week, and according to the United Nations (UN), by 2050, the urban population is expected to reach 6 billion. North America stands out as one of the most urbanized areas, with 82% of its population residing in cities (United Nations, 2014). Figure 1.1 clearly shows the decline in rural populations and simultaneous growth in urban populations.

Considering the information above, governments and relevant stakeholders are adopting innovative approaches and leveraging technologies to tackle a wide range of issues. However, the absence of coordination among urban policies, planning, and regulations results in underutilized urban spaces. This gives rise to

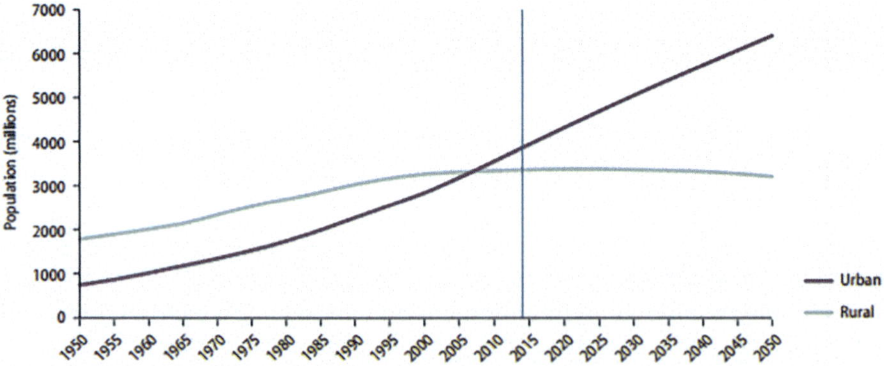

Fig. 1.1 Urban and rural population growth and projection. (Source: Data retrieved from un.org in Musa, 2017)

environmental ramifications, including heightened emissions of greenhouse gases and air pollution, which can be attributed to the scarcity of green or open spaces, which compels citizens to commute extensively to access services and commercial districts. Additionally, these haphazardly developed cities tend to foster social seclusion, marginalizing lower-income individuals and families who reside in poorly constructed housing and remain disconnected from essential urban amenities.

1.2 Transportation and Development

Defining development is a complex task, influenced by various schools of thought in the social sciences. Initially, development was perceived in narrow economic terms, such as an increase in per capita income during the 1950s and 1960s. However, the understanding of development has broadened over time (Okowa, 1991; Todaro, 1977). Okowa (1997) describes development as a process through which a society transforms in ways that benefit all or at least the majority of its citizens. Gboyega (2003) defines development as an effort to improve the conditions of human existence comprehensively, aiming to enhance the well-being of all citizens, eliminate or significantly reduce poverty and inequality of access to essential resources, and ensure sustainability. Development encompasses physical security, livelihoods, expanded life chances, and the advancement of society through strong partnerships among sectors, corporate entities, and various groups (Adeniran & Yusuf, 2016). It's important to recognize that development encompasses not only economic aspects but also socio-economic and political dimensions, permeating all aspects of society. Wilson (2005) defines development as the process of improving the overall welfare of a society, which is manifested through positive changes in various areas of life. These areas include reduced unemployment, decreased personal and regional inequalities, alleviation of absolute poverty, increased real output and improved production techniques, enhanced literacy, health services, housing, and government services, elevated social and political consciousness, better utilization of local resources to meet local needs, and reduced pollution or environmental degradation.

The term "transportation" derives from the Latin words "trans," meaning "across," and "portare," meaning "to carry." According to the Merriam Webster Dictionary, transportation is defined as the act, process, or instance of transporting or being transported. The Longman's Dictionary of Contemporary English (2003) and Good and Jebbin (2015) also describe it as a means of conveyance or travel from one place to another, including the commercial transport of passengers or goods using power-driven machines.

For transportation systems to function, they require several essential elements, including:

1. Infrastructure: This encompasses the transportation network, such as roads, railways, airways, canals, and pipelines.

2. Vehicles: These include automobiles, trains, airplanes, and other modes of transportation.
3. Operations: This involves the control and management of the transportation system, including traffic control, signaling, ramp meters, rail-road switches, air traffic control, and related policies such as tolls or gasoline taxes for highways.
4. Nodes or terminals: These are the points of connection within the transportation system, such as airports, railway stations, bus stations, and seaports.

Transportation is intricately linked to specific objectives and is not simply a standalone need. Its functions encompass spatial interaction, economic activities, and societal cohesion. Spatial interaction involves the transfer of individuals, commodities, or knowledge between different locations. It encompasses diverse forms of movement such as daily commuting, migration, tourism, utilization of public amenities, information dissemination, commercial transactions, international trade, and the distribution of goods (Rodrigue et al., 2006). Economic activities encompass the generation and attraction of various flows. The movement between different origins and destinations highlights the notion that the advantages gained from spatial interaction outweigh the associated costs. As an illustration, a commuter is willing to endure an hour-long drive due to the link with income, while international trade thrives on specialization and the resultant trade flows across distant locations (Rodrigue et al., 2006). Spatial interaction necessitates the fulfillment of three interconnected conditions:

- Complementarity: There must be a supply and demand relationship between the interacting locations. For instance, a residential zone supplies workers to an industrial zone, while a store and its customers or an industry and its suppliers (freight movements) exhibit complementarity. Reciprocity is common in commuting or international trade (Rodrigue et al., 2006).
- Intervening opportunity: There should be no closer alternative location that offers better opportunities as an origin or destination. The absence of a closer store offering similar goods, for example, ensures an interaction between a customer and a store.
- Transferability: The transfer of freight, people, or information must be supported by transport infrastructure, connecting the origin and destination. The costs of overcoming distance should not outweigh the benefits of the interaction. Adequate transport infrastructure, including modes and terminals, must be available to support the interaction (Rodrigue et al., 2006).

A study by Rodrigue and Notteboom (2020) emphasizes the crucial role of the transport sector in the economy and its significance for development. In a global economy, opportunities increasingly rely on the mobility of people, freight, and information, including communication technologies. The quantity and quality of transport infrastructure are closely linked to economic development. Regions with advanced and integrated transport networks tend to exhibit higher levels of development. Effective transport systems create economic and social opportunities, generating positive ripple effects such as enhanced market access, job creation, and greater investment. Conversely, inadequate transport systems in terms of capacity or reliability can result in economic costs, reduced opportunities, and a lower quality of life.

1.3 Transportation and National Development

Transportation's influence on the progress of a nation can be examined from diverse angles, encompassing economic, social, political, and environmental dimensions.

Economic Significance: Transportation exerts a substantial impact on national development, fostering economic growth through several avenues:

- Employment Generation: Transportation utilizes a substantial portion of the national labour force, serving as a source of employment for many individuals.
- GDP Growth: The transportation sector contributes significantly to the growth of a country's GDP by facilitating the movement of goods and services to both local and distant markets. This increased mobility leads to increased prices and higher profits for producers, encouraging businesses to expand their production capacity and output. Furthermore, efficient transportation services enable the smooth flow of production factors, specifically labour, capital, and entrepreneurship, fostering large-scale production, hence, promoting further economic advancement. For instance, countries like India, Nigeria and USA saw an increase in their GDP as a result of improvement in the transport sector. The transport sector in India is a significant contributor to the nation's GDP and serves as a crucial support system for various businesses and industries (Pathak et al., 2021). Projections indicate that by 2026, this sector will account for approximately 12% of India's GDP, up from the current 6.7%, positioning it as a major source of employment. Similarly, Placek (2023) noted that in 2019, the trucking industry had the highest impact on the United States' GDP within the transportation sector, contributing 368.9 billion U.S. dollars. According to the U.S. Department of Transportation, Bureau of Transportation Statistics (2021), the transportation sector added $1.3 trillion, or 5.6%, to the U.S. economy in 2021, up from 5.4% in 2020. In Nigeria, Ologunagbe (2022) highlighted that the road transport sector was one of the top performers in the third quarter of 2022, with a growth rate of 49.68%, based on data from the National Bureau of Statistics (NBS). These statistics underscore the vital role of the transportation sector in boosting a nation's economy, particularly through revenue generation.
- Geographical Specialization: Effective transportation services facilitate the geographical specialization of industries. By embracing geographical specialization, companies and regions can concentrate their efforts on producing goods and services in areas where they excel or face the fewest challenges. This strategic approach aligns with the Ricardian concept of comparative advantage and fosters efficient and optimized production processes. Without adequate transportation services, achieving specialization becomes challenging.

1.3.1 Social Impact of Transportation on National Development

Transportation services have a positive impact on the social life of a nation, enabling people to commute and maintain family or social interactions across different places and regions. For example, individuals residing in one part of a city can travel to other parts or even different cities, fostering social connections.

1.3.2 Political Impact of Transportation on National Development

The impact of transportation advancements on the political landscape of a nation can be viewed from two significant angles. Firstly, transportation plays a crucial role in the governance of states and the nation as a whole. Governments at different levels take the initiative to plan and establish viable transportation routes and infrastructure to ensure a well-functioning transportation network. This facilitates the smooth functioning of the government and enables effective governance. Secondly, transportation plays a key role in enhancing a nation's defense capabilities by providing cost-effective means to strategically deploy and maintain military forces at suitable locations. This aspect of transportation contributes to strengthening national defense and safeguarding the interests of the country.

1.3.3 Environmental Impact of Transportation on National Development

Environmental considerations are crucial in planning and operating transport infrastructure. It is essential to ensure environmental sustainability for the long-term viability of the transport industry. Many developed countries have made significant efforts in sustainable transport development through policies, measures, and innovations aimed at reducing environmental problems. However, the relationship between transportation and the environment is complex. While transportation activities cater to the escalating mobility needs in urban areas, they have also contributed to increased motorization and congestion, thereby exacerbating environmental issues. The transportation sector is now closely associated with significant environmental impacts, including climate change, air and noise pollution, water and soil quality degradation, loss of biodiversity, and land use alterations (Rodrigues et al., 2006; OECD, 1988).

1.4 Global Transportation Challenges

The impact of urban transport on the economy and individuals has been significant, but it has also presented challenges for the environment. Currently, addressing the issue of single occupancy vehicles (private vehicles) has become a priority for transportation planners worldwide. Stakeholders and governments have implemented measures aimed at reducing the number of cars on the roads to minimize the transportation challenges faced by many cities. The following problems associated with urban transport are highlighted:

1.4.1 Traffic Congestion

Traffic congestion arises when urban transport networks can no longer accommodate the volume of traffic. Globally, traffic congestion has worsened, posing a threat to the quality of urban life. Factors contributing to this include the rapid increase in motor vehicle ownership in developing countries due to improved purchasing power, reduced prices, and availability of automobiles. Population growth, smaller household sizes, and inadequate urban transport policies have also contributed to congestion (Cepal, 2003). The negative effects of congestion, such as longer travel times, increased fares, noise pollution, and environmental impacts, disproportionately affect lower-income individuals. Managing congestion is crucial for maintaining quality of life.

1.4.2 Inadequate Infrastructure

Many cities face challenges in providing adequate transportation infrastructure. Insufficient parking facilities lead to problems such as accidents and congestion as vehicle ownership grows. Parking demands consume significant amounts of space and inflate the demand for urban land. According to Paixoto Neto et al. (2008), at the time, infrastructure development has not kept pace with the growth in the number of vehicles, resulting in travel delays and capacity limits being reached. Large cities experience congestion throughout the day, hampering driving speeds and overall efficiency.

1.4.3 Public Transport Problems

Challenges and Limitations of Public Transportation in Urban Areas

- **Cost and Efficiency Concerns**

- As urban areas continue to experience spatial expansion, the financial burden associated with constructing and operating public transportation systems escalates. Public transport is widely regarded as an efficient mode of transportation for addressing urban issues due to its high capacity. Nonetheless, studies reveal a concerning trend of declining public transportation systems, especially in North America. The economic value of public transit is now under scrutiny due to the limited success of numerous urban transit initiatives in reducing congestion, despite significant expenses and considerable financial support (Cox, 1998).
- **Limited Availability and Convenience**
- Only approximately 80 major urban centers possess subway systems, with the majority situated in developed countries. Moreover, the dispersed residential layouts commonly found in car-centric cities make public transportation systems less accessible and user-friendly for facilitating urban mobility. As a result, despite additional investments in public transit in various cities, there has been minimal growth in ridership. The unplanned and disjointed development of land has led to rapid expansions on the outskirts of urban areas, where individuals choose to reside, limiting their ability to utilize public transportation effectively. Consequently, automobiles continue to dominate as the preferred mode of transportation within urban settings.
- **Government Dependence and Limited Economic Returns**
- Public transit systems are publicly owned, making them politically motivated services with limited economic profitability. Even in European cities, with a strong focus on transit-oriented development, public transportation systems heavily depend on government subsidies. These systems operate under strict regulations that restrict competition, leading to regulated wages and fares. This setup hinders the ability to adjust prices according to fluctuations in ridership. Consequently, while public transit serves as an important means of ensuring accessibility and promoting social equity, its connection to economic activities remains limited.

1.4.4 Lack of Integration of Transport Modes (Vehicle Mix)

One of the primary challenges in urban transport is the lack of comprehension regarding the roles of different transport modes, leading to a heavy reliance on a single mode. Road transport systems, railways, and even airlines need to collaborate and operate in an integrated manner to ensure sustainability. According to Masood et al. (2011), every mode, from walking and cycling to buses, forms part of the hierarchy of an urban transport system and should be optimally utilized. Moreover, pedestrian movement serves as the fundamental and oldest mode of transport, which remains relevant today. It is followed by bicycle use, motorcycles, and small vehicles such as Suzukis, van pickups, and trucks, with buses, trains, and planes being the last in the hierarchy. To achieve a balanced and cost-effective transport system, countries must plan for the integration of all transport modes.

1.4.5 Inadequate Funding

The perennial issue of inadequate funds to finance transportation projects has persistently plagued underdeveloped and developing countries. It is imperative to ensure the provision of sufficient financing for the construction of new projects and the maintenance of existing infrastructure, and the funding supply must be consistent and adequate. Consequently, exploring and implementing possibilities for increased cooperation between domestic and international private sectors is essential for planning and executing transportation projects.

1.4.6 Safety Concerns

Safety is a critical aspect of transportation that demands attention. Particularly in developing countries, the road safety situation has reached alarming levels. For example, Pakistan has become one of the countries with the highest accident rates globally. The major causes of accidents include inadequate signage, lack of driver education and training, and insufficient law enforcement both within and between cities.

1.5 The Need for Emerging Technologies

The world today is undergoing rapid transformation due to the swift progress and adoption of technology. This advancement in information and communication technology is compelling governments worldwide to transition from manual-based to information-based methods of service delivery (Al-Khouri, 2013). The term “emerging technologies“refers to science-based innovations that have the potential to create new industries or transform existing ones (Day & Schoemaker, 2000). These technologies can be described as either currently emerging or expected to be developed in the next few years or decades, significantly altering the social and business environment (Srinivasan, 2008). Vine (2002) further defined emerging technologies as those with the potential to gain social relevance within the next 10 to 15 years. Additionally, Haag, Cunnings, and James (1998) propose that emerging technologies fall into two categories: those that are so new that businesses have not yet exploited them, and those that are fairly well established but not fully utilized by businesses. These definitions demonstrate that emerging technologies are not necessarily brand new.

Nevertheless, these technologies serve as major drivers for the performance of public sector organizations, transforming government agencies, governance practices, and presenting new challenges for organizations and individuals (Desouza & Bhagwatwar, 2012). One of the most significant technological changes has been the evolution of information systems and their influence on technology usage worldwide. The combination of technology enabling the conversion of data into information has had a profound impact on the growth and development of urban environments and transportation in cities worldwide. These technologies have emerged as a result of remarkable progress in science and technology, significantly contributing to the field of urban transportation planning. In fact, emerging technologies themselves have arisen from revolutionary advancements in information and communications technology, producing various innovations that continuously enhance the governmental process and bring new features and attributes to the field of transport and urban planning. Consequently, emerging technologies have greatly enriched the scientific and practical dimensions of traffic management and urban environment management.

Emerging technologies are revolutionizing the transportation sector by introducing a wide range of advancements. These include the Internet of Things (IoT), drones, Geographic Information Systems (GIS), and Big Data, among others. These technologies have a dual impact, benefiting both commuters and governments. Commuters experience enhanced convenience and efficiency in their daily travels, while governments gain valuable insights to improve their transportation services and infrastructure. For example, by analyzing digital information obtained from swipe cards, the frequency of public transportation can be determined during specific hours of the day, assisting in better traffic management. Furthermore, police officers can navigate traffic and prevent congestion without physical presence, utilizing GPS tracking devices to analyze real-time traffic and manage traffic lights accordingly. The Internet of Things (IoT) can play a vital role in enhancing highway safety by facilitating communication between connected vehicles, thereby preventing accidents. Additionally, it has the potential to improve accessibility to quality healthcare through remote monitoring devices and telehealth practices, reducing response time delays and optimizing efficiency in factory supply chains and city operations. Smart cities not only bring benefits to citizens but also attract increased investments.

The utilization of technology, such as providing information on local attractions for tourism development, can contribute to a nation's GDP growth. These technologies work in tandem to collect and analyze vast amounts of data, enabling the enhancement of various components and systems within a city. The ICT framework, for example, comprises interconnected machines and sensors that transmit data instantly through wireless technology and the cloud. This allows for real-time monitoring of traffic in specific areas. Governments are actively transforming cities into "smarter" entities by harnessing data and digital technology to build more efficient and livable urban environments, stimulate economic growth, promote well-being, and encourage citizen engagement. As digitalization continues to advance, smart cities are increasingly recognized as unprecedented opportunities to drive economic growth and improve the well-being of residents in urban areas. To achieve the vision of smarter cities, governments need to foster the development of start-ups and industries that operate based on these technologies across various aspects of urban living.

1.6 Overview of the Global North and South

Today, we often simplify political and socioeconomic distinctions worldwide by dividing the globe into two categories: the Global South and the Global North. These terms are frequently used as synonyms for developing and developed countries. This is because most developing countries are situated in the southern hemisphere and include emerging economies and markets that typically have lower income levels and various structural challenges. Conversely, the Global North comprises developed economies and mature markets, primarily located in the northern hemisphere, characterized by high GDP per capita, advanced technology, robust infrastructure, and macroeconomic and political stability, among other attributes (Kowalski, 2020). Odeh (2010) reported that the terms "Global North" and "Global South" emerged after World War II in 1945, when many colonized nations gained independence, sparking significant scholarly interest in development. Colonialism had created a stark contrast between wealthy and impoverished societies, prompting scholars to explore comparative development. This led to the creation of terms such as "underdeveloped" and "developed," "metropole" and "satellite," and "center" and "periphery" to classify societies based on their economic development levels. "Developed," "metropole," and "center" referred to advanced economies, while "underdeveloped," "satellite," and "periphery" denoted less developed economies.

After the Cold War ended in 1991, the concepts of Global North and Global South were introduced in the comparative study of national development. The Global North refers to developed economies, characterized by wealth, democratic governance, peace, stability, and continuous progress. In contrast, the Global South is associated with turmoil, conflict, poverty, anarchy, and tyranny. Additionally, there is a significant disparity in institutional structures, such as democratic reforms, between the Global North and South. Historically, certain countries in the Global South, initially classified as developing, have advanced their economies and transitioned into developed nations. This shift is attributed to factors such as high levels of education, the establishment of global enterprises, and technological innovation. Examples of such transformations include the Asian Tigers—South Korea, Singapore, Taiwan, and Hong Kong—known for their competitiveness and innovation. Today, the Global South predominantly includes regions in Latin America, Africa, and parts of developing Asia.

1.6.1 Concept of the Global North and South

In the 1940s, the United Nations established the Three-World model as a way to classify countries according to their political systems. The First World consisted of nations allied with the United States, distinguished by their capitalist economies and political and economic stability. In contrast, the Second World included countries aligned with the Soviet Union, which operated under socialist or communist systems. The Third World encompassed countries not aligned with either the US or the Soviet Union, primarily located in Latin America, Africa, Oceania, and Southeast Asia. Due to criticism from Third World member states regarding this term, it was eventually replaced with "developing economies" or "the Global South." Similarly, First World countries are now referred to as developed economies, and Second World countries are known as transition economies. Although the term "Third World" is no longer in use, the global hierarchy remains highly stratified, with many countries outside the Northern core exhibiting similar positions within the international system, despite significant local differences (Lees, 2011). One major challenge for the development of Global South countries is their reliance on agriculture and the export of raw materials, whose prices are highly volatile, negatively impacting their economies. These countries often have underdeveloped industries with insufficient capital for investment in machinery and technology. Despite being rich in natural resources, the Global South faces the "Natural Resource Curse" or "paradox of plenty," where resource-rich economies tend to have lower income levels, higher corruption, and poorer development outcomes compared to resource-scarce countries (Sachs & Warner, 2001). The classification of countries into the Global South and Global North is shown in Table 1.1.

Income differentials are a key distinguishing factor between countries in the Global North and Global South. The World Bank classifies countries as developed or developing based on Gross National Income (GNI) per capita. According to the World Bank's latest database, a developed country in 2019 is defined as having a GNI per capita of $12,056 or more, while a developing country has a GNI per capita of $12,055 or less. Within the developing countries, there are further classifications: upper-middle-income economies (GNI per capita from $3896 to $12,055), lower-middle-income economies (GNI per capita from $996 to $3895), and low-income economies (GNI per capita of $995 or less). The regions leading in GDP per capita are predominantly part of the Global North, including North America and Europe. Conversely, regions in the Global South have GDP per capita figures below the world average of $10,634 in 2017. These regions include Sub-Saharan Africa ($1648), South Asia ($1779), the Middle East and North Africa ($7740), Latin America and the Caribbean ($9356), and East Asia and the Pacific ($9915) (Kowalski, 2020).

Technological advancement and Innovative factors is another criteria used in distinguishing between the global north and south. Various economic theories, including the new theory of growth, new economic geography, and the concept of a knowledge-based economy, highlight innovation as a crucial factor for international competitiveness. Historically, technological progress driven by research and development has been concentrated in developed countries, which have been responsible for most of the innovation seen in global development (Furman & Hayes, 2004; Dosi et al., 2006). Consequently, the competitive advantage of Global North countries in recent decades has been linked to factors such as knowledge and innovation, positioning these countries as technological leaders in the world economy. In contrast, the Global South has primarily relied on cheap resources, leading to specialization in labor-intensive industries. Theoretically, technological disparities between

Table 1.1 Classifications of the countries in the Global South and Global North

Region	Countries
The Global South	
Latin America and Caribbean	Antigua and Barbuda, Aruba, Argentina, Bahamas, The Barbados, Belize, Bolivia, Brazil, Cayman Islands, Chile, Costa Rica, Colombia, Cuba, Curac¸ao, Dominica, Dominican Republic, Ecuador, El, Salvador, Grenada, Guatemala, Guyana, Haiti, Honduras, Jamaica, Mexico, Nicaragua, Panama, Paraguay, Peru, Puerto Rico, St. Kitts and Nevis, St. Martin, St. Lucia, St. Vincent and the, Grenadines, Suriname, Trinidad and Tobago, Turks and Caicos Islands, Uruguay, Venezuela, Virgin Islands
East Asia and Pacific	American Samoa, Brunei, Darussalam, Cambodia, China, Fiji, French Polynesia, Guam, China, Indonesia, Kiribati, North Korea, Lao PDR, Malaysia, Marshall Islands, Mongolia, Myanmar, Nauru, New Caledonia, Northern Mariana Islands, Palau, Papua New Guinea, Philippines, Samoa, Solomon Islands, Thailand, Timor-Leste, Tonga, Tuvalu, Vanuatu, Vietnam
Central Asia	Azerbaijan, Kazakhstan, Kyrgyzstan, Tajikistan, Turkmenistan, Uzbekistan
South Asia	Afghanistan, Bangladesh, Bhutan, India, Pakistan, Nepal, Maldives, Sri Lanka
Middle East and North Africa	Algeria, Bahrain, Egypt, Djibouti, Iraq, Iran Islamic, Rep., Jordan, Kuwait, Lebanon, Libya, Malta, Morocco, Oman, Qatar, Saudi Arabia, Syrian Arab Republic, United Arab Emirates, Tunisia, Yemen
Sub-Saharan Africa	Angola, Benin, Botswana, Burkina Faso, Burundi, Cabo Verde, Cameroon, Central African Republic, Chad, Comoros, Congo Dem. Rep., Congo Rep., Côte d'Ivoire, Ethiopia, Eritrea, Equatorial Guinea, Gabon, Gambia, The Ghana, Guinea, Guinea-Bissau, Kenya, Lesotho, Liberia, Madagascar, Malawi, Mali, Mauritania, Mauritius, Mozambique, Namibia, Niger, Nigeria, Rwanda, São Tomé and Principe, Seychelles, Senegal, Sierra Leone, Somalia, South Africa, South Sudan, Sudan, Eswatini, Tanzania, Togo, Uganda, Zambia, Zimbabwe
The Global North	
Europe	Albania, Andorra, Armenia, Austria, Belarus, Belgium, Bosnia and Herzegovina, Bulgaria, Croatia, Cyprus, Czech Republic, Denmark, Estonia, Faroe Islands, Finland, France, Georgia, Germany, Gibraltar, Greece, Greenland, Hungary, Iceland, Ireland, Isle of Man, Italy, Latvia, Liechtenstein, Lithuania, Luxembourg, Macedonia FYR, Moldova, Monaco, Montenegro, Netherlands, Norway, Poland, Portugal, Romania, Russian Federation, San Marino, Serbia, Slovak Republic, Slovenia, Spain, Sweden, Switzerland, Turkey, Ukraine, the United Kingdom
North America	Rep. Bermuda, Canada, the United States
Other	Exceptions from East Asia & Pacific: Australia, Hong Kong, Japan, Macau, New Zealand, Singapore, South Korea, Taiwan Exception from Middle East & North Africa: Israel

Source: World Bank classification in Kowalski (2020)

countries provide an opportunity for those with lower innovation profiles to catch up by imitating the more productive technologies used in leading countries, as imitation is generally less costly than innovation. Based on the above, the role of technology in defining a country as either developed or developing cannot be over emphasized due to the fact that technological advancement opens up a country to a world of innovation, development and productivity.

From Table 1.1, it can be deduced that different countries based on their varying characteristics such as level of technological achievement and income differentials among other factors were used to group them into the two categories. This book hence selected three countries each from the global south and global north based on geographical classification. For the global south, United States of America was selected, Germany from Europe and Singapore from Asia. Likewise in the global South, two countries from the sub-Saharan Africa was selected and these are Nigeria from West Africa and South Africa, in order to choose a representative sample from Asia, India was selected for review in the book.

1.7 Aim and Objectives

This book examines the need for technological adoption in the transportation planning of our cities in both developed and developing countries of the world and analyses the roles of government and relevant stakeholders in enhancing sustainable cities through the use and adoption of emerging technologies.

The objectives of the book are summarized as follows:

(a) To identify and examine the various transportation problems in cities of both developed and developing countries of the world.
(b) To explore the various transportation technologies and their applications in transportation planning.
(c) To examine the benefits of these technologies in achieving sustainable growth of the urban environment.
(d) To identify the various challenges militating against the efficient utilization and adoption of emerging technologies in cities of both developing and developed countries.
(e) To identify the roles of different stakeholders, government and citizens in partnering to ensuring sustainable growth and development through the adoption of emerging technologies.

1.8 The Contribution and Value of Emerging Technologies in Urban Planning

- **Bridging the Gap and Examining Transport Situations**

This book serves as a highly innovative and timely resource for the built environment and academia, aiming to promote the adoption of emerging technologies for effective city management and planning. Its contributions are multifaceted and significant. To begin with, this book fills a significant void by exploring the distinct transportation challenges found in both advanced and emerging nations, a topic that has garnered substantial scholarly attention. Additionally, the book underscores the significance of embracing a comprehensive, multimodal transportation strategy as a pathway toward sustainable development. By providing practical insights and solutions, it advocates for reducing reliance on single occupancy vehicles, thereby mitigating traffic congestion and other transportation-related issues in urban areas.

- **Technologies for Smart and Sustainable Cities**

Another vital contribution of this book is its exploration of various technologies that can be adopted to enhance urban management and transform cities into smart cities. It examines the role of each technology in improving the mobility of city dwellers while addressing environmental concerns such as greenhouse gas emissions and ozone layer depletion. By identifying these technologies, the book offers valuable insights into creating more sustainable and environmentally friendly urban environments.

- **Fostering Collaboration and Skill Development**

Furthermore, the book aims to address the lack of collaboration among stakeholders in the built environment, government, and citizens. It emphasizes the need for collective efforts in developing the necessary skills for transitioning to modern technologies. These technologies provide real-time information on traffic conditions and enable proper planning of transport systems. By fostering collaboration and knowledge-sharing, the book seeks to facilitate the epochal transition towards utilizing emerging technologies effectively.

- **Achievement of some of the agendas of the Sustainable Development Goals** (SDGs) **and New Urban Agenda**

The Sustainable Development Goals (SDGs) advocate for initiatives to ensure peace and prosperity for all individuals and the planet, extending from the current era to future generations. These goals primarily focus on reducing inequalities in health, education, and the economy, while also addressing climate change and protecting marine environments. This book aims to enhance understanding of these objectives, particularly emphasizing Goal **11**, which is centered on creating sustainable cities and communities. This goal seeks to make urban areas and human settlements safer, more resilient, and sustainable by implementing national urban policies, increasing access to public spaces, and providing safe and convenient public transportation. Technology plays a crucial role in offering innovative solutions for comprehensive city management, including ICT infrastructure and applications like smart buildings, intelligent transportation systems, and improvements in energy consumption and waste management (Adebayo, **2018**). The concept of smart cities involves utilizing ICT to manage urban resources more effectively. Oluwayemi-Oniya Aderibigbe & Trynos Gumbo (**2024**)), asserted that the impact of ICT tools such as intelligence transport system in this aspect will contribute to improvements in traffic planning. For instance, the discovery and adoption of electric and autonomous vehicles being one of the current innovations in the transport sector highlights the impact of ICT at combating some of the environmental problems accrue from the reliance on vehicles that use fossil fuels, hence, endangering the health of city inhabitants and the environment at large.

1.9 Summary

This chapter provides an extensive introduction to the concept of smart and sustainable cities, emphasizing the role of emerging technologies in sustainable planning. It explores the impact of transportation on economic growth and development and examines the transport challenges faced by cities in both developed and developing countries. Additionally, the chapter highlights the objectives and values contributed by the book in addressing these challenges and promoting the adoption of emerging technologies in urban planning.

1.10 Structure of the Book

This book is divided into three parts which are highlighted below:

Part 1 which addresses the introduction and Setting the Scene: This part consists of three chapters which provide a general overview of the book as well as the transport situation in the countries of the Global North and Global South. The chapters in this part are: Chapter 1: Emerging technologies and concept of transportation and development. Chapter 2 which highlights transportation development in the global north: lessons from Germany, United States of America and Singapore while Chap. 3 addressed the transport situation in the global south: evidence from Nigeria, South Africa and India. Part II of the book discussed Sustainable Development, Smart Cities and Emerging Technologies. This part has four chapters which highlighted the issues of sustainable planning and different technologies with their application which are beneficial in urban and transportation planning. The four chapters contained in this part are Chap. 4, Sustainable Transport Planning, Chap. 5, smart cities and their impact on urban transportation systems and development, Chap. 6 examined transportation technologies and transportation management and Chap. 7 addressed Internet of Things (IoT) and Big Data as a Revolutionary Approach for traffic management. Finally, the last part of the book considered prospects of emerging technologies and the role of government and citizen in smart cities and urban transportation development has two chapters namely Chap. 8 which discussed the benefits and challenges of emerging technologies in urban transportation planning and Chap. 9 addressed smart technologies and smart cities: the role of public and community participation in governance.

References

Adebayo, T. J. (2018). *The role of technology in achieving the SDGs 2030.* https://www.academia.edu/38900809/THE_ROLE_OF_TECHNOLOGY_IN_ACHIEVING_THE_SDGs_2030. Assessed 12 Jan 2024.

Adeniran, A. O., & Yusuf, T. B. (2016). Transportation and National Development: Emphasis to Nigeria. *Developing Country Studies, 7*(9), 93–104.

Al-Khouri, A. M. (2013). Technological and mobility trends in e-government. *Business and Management Research, 2*(3), 90–113.

Cepal. (2003). *Traffic congestion: The problem and how to deal with it*. https://repositorio.cepal.org/bitstream/handle/11362/37898/1/LCG2199P_en.pdf

Cox, W. (1998). *Light rail in Minneapolis: A bridge to nowhere, the public purpose*. Urban Transport Fact Book.

Day, G., & Schoemaker, P. (2000). *Wharton on managing emerging technologies*. Wiley.

Desouza, K. C., & Bhagwatwar, A. (2012). Leveraging technologies in public agencies: The case of the U.S. Census Bureau and the 2010 census. *Public Administration Review, 72*(4), 605–614.

Dosi, G., Llerena, P., & Sylos Labini, M. (2006). The relationships between science, technologies and their industrial exploitation: An illustration through the myths and realities of the so-called 'European Paradox'. *Research Policy, 35*, 1450–1464.

Furman, J. L., & Hayes, R. (2004). Catching up or standing still?: National innovative productivity among "follower" countries, 1978–1999. *Research Policy, 33*, 1329–1354.

Gboyega, A. (2003). *Democracy and development: The imperative of local governance. An inaugural lecture*. University of Ibadan.

Good, W., & Jebbin, M. F. (2015). Transportation and National Development. *Journal of Economics and Sustainable Development*. www.iiste.org ISSN 2222-1700 (Paper) ISSN 2222–2855 (Online).6,(9), 2015.

Haag, S., Cunnings, M., & James, D. (1998). *Management information systems for the information age*. Irwin McGraw-Hill.

Kim, M., Lim, C., Lee, C., Kim, K., Park, Y. & Choi, S. (2018). Approach to service design based on customer behaviour data: A case study on eco-driving service design using bus driver's behaviour data. *Service Business, 12*(1), 203–227.

Kowalski, A. M. (2020). Global South-Global North differences. In W. Leal Filho, A. Azul, L. Brandli, A. Lange Salvia, P. Özuyar, & T. Wall (Eds.), *No poverty. Encyclopedia of the UN Sustainable Development Goals*. Springer. https://doi.org/10.1007/978-3-319-69625-6_68-1

Lawal, T., & Abe, O. (2011). National development in Nigeria: Issues, challenges and prospects. *Journal of Public Administration and Policy Research, 3*(9), 237–241., November 2011 online http://www.academicjournals.org/jpapr. ISSN 2141-2480. https://doi.org/10.5897/JPAPR11.012

Lees, N. (2011). The dimensions of the divide: Theorising inequality and the Brandt Line in international relations. Paper presented as part of the Panel "South South Cooperation: History, Concepts, Trends" In: *IPSA-ECPR Joint conference 'Whatever happened to North-South?'*, Sao Paolo.

Longman Dictionary of Contemporary English. (2003). (4th ed.). Pearson Education Limited.

Masood, M. T., Khan, A., & Naqvi, H. A. (2011). Transportation problems in developing countries Pakistan: A case-in- point. *International Journal of Business and Management, 6*(11), 256–266. https://doi.org/10.5539/ijbm.v6n11p256

Merriam Webster Dictionary. (2012). Inc.

Mitchell, M. (2007). Intelligent city. UOC Papers. *e-Journal of the Knowledge Society*. Retrieved from: https://pdfs.semanticscholar.org/6c8c/d3f7e497c7ee75c6c54c737e84cec5f78418.pdf. 21.10.2019

Musa, S. (2017). Smart cities: Recent trends and future direction in Malaysia. *Journal of Information Systems Research and Innovation, 9*(1), 39–46.

Odeh, L. E. (2010). A comparative analysis of Global North and Global South economies. *Journal of Sustainable Development in Africa, 12*(3), 338–348. ISSN:1520-5509.

Okowa, W. J. (1991). *The political economy of development planning in Nigeria*. Paragraphics Publishers.

Okowa, W. J. (1997). *Oil, systemic corruption, abdulistic capitalism and Nigerian Development Policy: A political economy*. Paragraphics Publishers.

Ologunagbe. (2022). Road transport ranks high in Q3 2022 GDP. *Business Day report*. https://businessday.ng/transport/article/road-transport-ranks-high-in-q3-2022-gdp/#:~:text=quarter%20of%202022.-,The%20contribution%20of%20the%20sector%20to%20real%20GDP%20in%20the,2022%20year%2Don%2Dyear. Assessed 17 Jan 2021.

Oluwayemi-Oniya Aderibigbe & Trynos Gumbo. (2024). Impact of information and communication technology on mobility of urban and rural households: A comparative study from Nigeria. *Urban, Planning and Transport Research, 12*(1), 2348551. https://doi.org/10.1080/21650020.2024.2348551

Organization for Economic Co-operation and Development (OECD). (1988). *Transport and the environment*. OECD.

Paixoto Neto, A. G. L., Galves, M. L., Lima Junior, O. F., & Tacla, D. (2008). Challenges of urban transport problems and city logistics: Sao Paulo City Center Case. *Wits Transaction on the Built Environment, 101*, 133–142.

Pathak, P., Singhal, A., & Bhutani, S. (2021). Intelligent Transportation System in India. *Telecom Business Review: SITM Journal*. file:///Users/macbook/Downloads/Intelligent_Transportation_System_in_India20190923-80370-usf15h-with-cover-page-v2.pdf. Assessed 22 Oct 2023.

Placek, M. (2023). GDP of the U.S. attributed to transportation by mode and type of transportation 2019. *Statista*. https://www.statista.com/statistics/1342180/united-states-gdp-attributed-to-transportation-by-mode-type/. Assessed Dec 2023.

Rodrigue, J. P., & Notteboom, T. (2020). Transportation and economic development. In *The geography of transport system* (5th ed.). Published Routledge. https://doi.org/10.4324/9780429346323

Rodrigue, J. P., Claude, C., & Brian, S. (2006). *Geography of transport systems*. First published in 2006 by Routledge.

Sachs, J. D., & Warner, A. M. (2001). The curse of natural resources. *European Economic Review, 45*(4–6), 827–838.

Srinivasan, R. (2008). Sources, characteristics and effects of emerging technologies: Research opportunities in innovation. *Industrial Marketing Management, 37*, 633–640.

Todaro, M. P. (1977). *Economic Development in the Third world: An introduction to problems and policies in a global perspective*. Longman, London.

U.S. Department of Transportation, Bureau of Transportation Statistics. (2021). *Transportation satellite accounts*. Available at www.bts.gov/satellite-accounts

Vine, E. (2002). Promoting emerging energy-efficiency technologies, and practices by utilities in a restructured energy industry: A report from California. *Energy, 27*(4), 317–328.

Wilson, G. (2005). Chapter 5: Economic and social infrastructure: Provision, distribution and maintenance. In C. Amadi & N. Amadi (Eds.), *Applied economics in perspectives* (pp. 93–128). ROD Printing and Publishing Company.

Chapter 2
Transportation Development in the Global North: Lessons from Germany, United States of America and Singapore

Oluwayemi-Oniya Aderibigbe

2.1 Introduction

Transportation involves the movement of people, goods, and information from their origin to their destination. According to Rodrigue et al. (2012), these movements have always been integral to human societies, and with contemporary economic processes, there has been a substantial upsurge in both mobility and accessibility. While this trend can be traced back to the industrial revolution, it experienced a notable acceleration during the latter part of the twentieth century, primarily driven by the liberalization of trade. The primary purpose of transportation is to overcome spatial challenges influenced by various human and physical factors, such as distance, time, administrative divisions, and topography. These factors collectively create friction, impeding movement through space. However, it is only possible to partially circumvent these constraints and the resulting friction. The achievement of this goal comes with associated expenses that differ based on factors such as distance, transportation modes, infrastructure capacity, and the characteristics of the goods being transported. The aim of transportation is thus to effectively transfer freight, individuals, or information from their point of origin to their intended destination, while simultaneously enhancing their value throughout the journey. Transportation infrastructure, as a complex network, connects cities and facilitates human activities, effectively integrating social, economic, and environmental systems with urbanization and population growth. Moreover, the transportation network contributes to socioeconomic development and an improved quality of life by establishing inter- and intra-city connections during urbanization. Considering the aforementioned factors, the significance of transportation in both developed and developing countries cannot be overstated. However, despite its numerous benefits, transportation development has resulted in severe problems, particularly in urban areas where reliance on motorized means of transportation is high. In light of these

O.-O. Aderibigbe et al., *Emerging Technologies for Smart Cities*,
https://doi.org/10.1007/978-3-031-66943-9_2

challenges, this chapter examines the transportation situation in select developed countries worldwide, aiming to understand the measures and policies implemented to enhance their transportation systems.

2.2 An Overview of Transportation Situations in the Global North: The Cases of Germany, the United States of America and Singapore

In recent years, there has been increasing recognition in Europe and North America about the importance of enhancing the social, environmental, and economic sustainability of transportation systems (Banister, 2005). However, many countries, including the USA, have struggled to achieve their transport sustainability goals (Newman & Kenworthy, 1999; Vuchic, 1999; Low & Gleeson, 2002; Tolley, 2003). The private car has become the dominant mode of transportation, primarily due to its high levels of mobility and convenience, as demonstrated by numerous studies (Ingram & Liu, 1999; Schafer & Victor, 2000; Downs, 2004). Increased income and improved affordability of cars have contributed significantly to the global motorization trend observed in recent decades. However, as car usage continues to rise, it tends to displace and limit the availability of other travel options. This phenomenon stems not only from the inherent benefits of car use but also from the negative impacts it has on alternative modes of transportation. For instance, cars pose significant risks to pedestrians and cyclists, leading to a decline in walking and cycling (Tolley, 2003; FHWA, 2004, 2009; IRTAD, 2008). Additionally, car congestion on roadways hampers the efficiency of buses and discourages public transport use (Downs, 2004; Vuchic, 2005). Buehler and Pucher (2009) note that cars also contribute to low-density, sprawling development, which is challenging to serve with public transport and generates trip distances that are impractical for walking or cycling. Consequently, the increasing reliance on cars diminishes the attractiveness of sustainable modes of transportation, resulting in a modal shift towards car use (Newman et al., 1999; Banister, 2005). As a result, public transport, walking, and cycling have experienced a decline in market share across many cities worldwide (Newman et al., 1999; Banister, 2005). It is crucial for policymakers and researchers to address these challenges and promote sustainable alternatives to car-centric transportation systems.

2.3 Transport Situation in Germany

Germany, with a population of 82 million people and 43 million cars, exhibits a high degree of motorization compared to other European countries. Over the years, the share of cars and motorbikes in motorized transport has increased significantly,

Fig. 2.1 An overview of night traffic on a typical German freeway. (Source: Wendt, 2018)

accounting for 73% of the total volume in 1999, up from 64% in 1960 (Verron & Friedrich, 2004). This trend has led to a high dependence on cars in Germany, resulting in various transportation-related issues. To address these challenges and reduce car dependency, the German government has implemented a long-term strategy that incorporates digital mobility services, innovative mobility solutions, and car sharing into the framework of autonomous and connected driving for public transport. This approach aims to foster sustainable transportation planning and mobility (Hampel, 2021). Over the years, German governments at all levels have played a crucial role in shaping travel behavior through the implementation of policies targeting car use. These policies encompass pricing mechanisms, restrictions, and technological improvements aimed at mitigating the negative impacts of car dependency (Buehler et al., 2009). Additionally, efforts have been made to improve the integration of public transportation systems at both the metropolitan and national levels, providing viable alternatives to car travel. Furthermore, Germany has made significant strides in land-use planning and road infrastructure development (see Fig. 2.1). Regional land planning policies have been implemented to encourage compact, mixed-use development, resulting in shorter trip distances and promoting walking or cycling as feasible modes of transportation. These comprehensive measures and policies adopted by Germany exemplify a concerted effort to address the challenges associated with high car dependency, promote sustainable transportation planning, and enhance mobility options for its citizens.

2.4 Overview of German Transport and Policies Aimed at Sustainable Transportation

Germany has implemented a range of strategies to foster sustainable transportation practices. Firstly, there are regulations in place to limit car usage and reduce its negative impact. Secondly, the country has prioritized the development of efficient, affordable, and well-coordinated public transportation systems, especially in major urban areas, as a viable alternative to cars. Thirdly, substantial investments have been made in improving transportation infrastructure to support non-motorized modes of travel, making walking and cycling safer and more convenient. Lastly, urban planning and land use policies prioritize compact and mixed-use developments, discouraging sprawling suburbs and ensuring that travel distances remain short enough to encourage walking and cycling as convenient transportation options. These comprehensive efforts demonstrate Germany's commitment to promoting sustainable transportation practices.

2.4.1 Pricing and Restrictions on Car Use

According to Buehler and Pucher (2009), the total cost of owning and operating a car in Germany is approximately 50% higher compared to other developed countries. This disparity can be attributed largely to the significantly higher taxes and fees imposed on car ownership and usage in Germany. For instance, fuel taxes in Germany were nine times higher in 2006 than in other developed regions. These taxes and fees have seen an increase over time, with the resulting revenues being allocated to various government expenditures, including road construction and maintenance. Moreover, German cities have implemented measures to limit car usage, such as restricted road supply, lower speed limits, and reduced parking availability. Despite having the third-largest motorway network globally, German motorways rarely extend into city centers, effectively reducing traffic congestion in urban areas (Pucher, 1995). Germany has implemented various strategies to discourage over reliance on cars. These strategies include the creation of extensive car-free zones, deliberate placement of dead-ends, implementation of turn restrictions, and establishment of one-way street networks. These measures are designed to make it difficult for vehicles to pass through the city center, forcing alternative modes of transportation (Pucher, 1988; Hajdu, 1989; Hass-Klau, 1993; Topp, 1993). Additionally, lower speed limits have been enforced, ensuring that vehicles travel at a pedestrian-friendly pace. For instance, a significant portion, approximately 70–80%, of the road network in German cities and small towns have speed limits of 30 km/h or less (Beatley, 2000; Newman et al., 2009).

2.4.2 Public Transport Improvements

Germany is renowned for its exceptional transportation services, surpassing other European countries and many developed nations worldwide. The country boasts a comprehensive, high-quality, and well-integrated public transport network. Despite significant increases in per-capita income and car ownership, the utilization of public transport has continued to rise. For instance, between 1970 and 2005, public transport trips per capita increased from 116 to 133 in Germany (Buehler & Pucher, 2009). One of the key reasons for the economic sustainability of German public transport is the higher passenger fare revenues coupled with lower costs. German buses, trams, metros, and trains typically accommodate more than twice the number of passengers compared to other countries. Additionally, the vehicles are relatively new and increasingly reliable, which further attracts passengers. The success of German public transport can also be attributed to the coordination and integration of various modes of transportation, fares, and schedules within metropolitan areas. For example, Hamburg city has achieved full integration of all aspects of public transport operations and financing (Pucher & Kurth, 1996). Similar integrated public transport organizations have been established in almost all metropolitan areas across Germany, resulting in expanded and improved services, enhanced fare structures, and substantial increases in passenger numbers.

The German public transport systems have effectively facilitated seamless transfers for passengers, both in terms of timing and distance walked, enabling easy transitions between different modes of transport, routes, and operators. Moreover, these systems offer significant discounts on weekly, monthly, annual, and semester tickets, making public transport economically competitive with private cars for daily commuting purposes (VDV, 2005). Another noteworthy aspect of German public transport is its successful integration with walking and cycling facilities. Well-maintained sidewalks, safe pedestrian crossings, and car-free zones ensure easy access to bus and rail stops for pedestrians. Additionally, extensive bicycle parking facilities are provided by almost all German public transport systems (Pucher & Buehler, 2008). Therefore, the success of public transport in Germany can be attributed not only to financial investments but also to effective utilization of subsidies, well-designed fare structures, and comprehensive service policies. These factors contribute to the overall efficiency, sustainability, and attractiveness of the German public transport system.

2.4.3 Active Travel Use in Germany

Transport infrastructure provision and maintenance, specifically focusing on non-motorized transport, has witnessed significant improvements. German cities, for instance, have made substantial advancements in enhancing non-motorized transport infrastructure for pedestrians and bicyclists (BMVBS, 2000). The city centers are encompassed by free zones, and there are spacious sidewalks on both sides of every street, catering to pedestrians. Design features that promote pedestrian safety include pedestrian refuge islands for crossing wide streets, well-marked zebra crosswalks (often raised) with special lighting for improved visibility, and pedestrian-activated crossing signals (Pucher & Dijkstra, 2003). Residential and commercial areas incorporate sidewalks for pedestrians, and many developments provide separate bike paths along with ample bicycle parking facilities. The bicycling and walking networks in German cities are comprehensive, featuring off-street shortcut connections (refer to Fig. 2.2) that enable cyclists and pedestrians to choose the most direct routes from their starting point to their destination. This integrated system comprises segregated paths and lanes, as well as lightly trafficked, residential streets with traffic-calming measures, ensuring minimal interaction with cars (Pucher & Buehler, 2008). As a result, there is a reduction in traffic congestion due to fewer vehicles on the road, a decrease in accidents, and an environment free from pollution caused by excessive car reliance. The commendable pedestrian and cycling safety in Germany can be largely attributed to various government policies aimed at promoting it. These policies encompass extensive networks of bike paths and lanes on busy arterial streets, the implementation of priority traffic signals at intersections, and comprehensive traffic calming measures in residential neighborhoods, all of which enhance pedestrian safety (Fig. 2.3).

Fig. 2.2 An overview of a pedestrian walkway and cycle path for active travel in Germany. (Source: Adriazola-Steil and Schwedhelm, 2020)

Fig. 2.3 Bicycle path in Berlin, Germany. (Source: Bicycle Dutch, 2013)

2.4.4 Urban Development and Land Use Policies

Cities in Germany have undergone a process of decentralization (Nivola, 1999; Burchell et al., 2002; Divall & Bond, 2003). However, it is worth noting that as of 2003, the average population density in German cities and suburbs was up to three times higher compared to the USA. This higher density, combined with a greater mix of land uses, resulted in shorter average trip distances, thus facilitating walking and cycling opportunities. The increased population densities also made public transport services more economically viable by generating higher passenger volumes. In Germany, the land-use planning process involves cooperation, compromise, and mediation among the federal, state, regional, and local governments (Kunzmann, 2001). Moreover, formal links exist at each government level between land use and other planning areas such as transport and the environment, promoting coordination (see Fig. 2.4) (Fuerst & Scholles, 2003). This coordination is facilitated by reduced competition among municipalities for property taxes. According to Schmidt and Buehler (2007), the achievement of effective and well-integrated compact and mixed-use development in Germany can be attributed to the collaborative efforts between jurisdictions at the same governmental level and the cooperative relationships established between different levels of government. Furthermore, the implementation of stringent regulations at the outskirts of suburban areas promotes the establishment of high-density and mixed-use developments. Additionally, the presence of tax-sharing arrangements plays a crucial role in minimizing competition among cities and towns for the tax base.

From the aforementioned studies, it can be concluded that Germany is among the developed nations that prioritize sustainable transport planning. This is evident in

Fig. 2.4 An overview of mixed land use in Berlin Germany (A mix of residential area easily accessible to essential facilities (road network, shopping, recreational). (Source: Sharma. https://www.quora.com/What-is-mixed-land-use)

the formulation and implementation of policies aimed at promoting non-motorized transport and reducing reliance on private automobiles. Overall, Germany has made significant improvements in its transportation infrastructure, providing quality and improved mobility options for residents, thereby enhancing their quality of life.

2.5 Transport Situation in the United States of America

The United States has traditionally been known for its high private car use and ownership. Bormans (2018) reported that there are approximately 766 cars per 1000 people, with around 70% of the population possessing driver's licenses, indicating a heavy reliance on private vehicles. However, various factors such as changes in demographics, technology, economics, and cultural preferences have led to a gradual decline in private car usage. Evidence suggests that the use of private vehicles has been slowly decreasing over time. Martin et al. (2016) and Polzin (2018) conducted studies revealing the following trends:

- The amount of travel by vehicle (VMT) has remained relatively steady since the economic downturn known as the Great Recession.
- The average distance driven by Americans has decreased by 7.6% compared to 10 years ago.
- There has been a slight decline of 0.4% in the demand for new cars.
- The average number of daily trips taken has decreased from 4.3 in 1995 to 3.4 in 2017.
- The proportion of daily commutes made in single-occupancy vehicles (SOV) has seen a small decrease from 92.8% in 2001 to 91.4% in 2009.

Additionally, there has been a decline in car ownership among Americans. Bouton et al. (2015) reported a 5–6% decrease in the number of vehicles per person, per driver, and per household from 2006 to 2012, with the millennial generation experiencing the most significant declines. Martin et al. (2016) also observed an increase in the percentage of households with zero to one car and a decrease in households with two or more cars. Urbanization and advancements in internet technology have further contributed to the reduction in private car use and ownership. People residing in densely populated urban areas are more likely to own fewer vehicles. Additionally, the rise of e-commerce has led to fewer shopping trips, while telecommuting and remote-work capabilities have allowed more individuals to work from home, resulting in decreased reliance on private cars. In conclusion, several factors, including demographics, locational attributes, and information and communication technologies, have contributed to the decrease in private car use in the United States (Fig. 2.5).

2.6 Overview of United States Transport Policies Aimed at Sustainable Planning

2.6.1 *Demographic Changes*

The aging of the American population is a significant demographic trend, with projections indicating that by 2030, one in five Americans will be 65 years or older (Martin et al., 2016; Borade et al., 2020). In terms of generational impact on American mobility, the millennial generation (born 1981–1996) has now surpassed the baby boomers (born 1946–1964) as the largest adult generation in the United States. Consequently, the travel behaviors of millennials will have the most substantial influence on future American mobility. Notably, car ownership rates have experienced the greatest decline among millennials compared to other generational cohorts. Moreover, there has been a significant decrease in the number of young people obtaining driver's licenses, reaching an all-time low. This decline in license

Fig. 2.5 Road network with designated lanes for Buses, bicycles and pedestrians to facilitate utilization of different modes of transport in the United States of America. (Source: Shyshkina 2020)

rates serves as an indication that car ownership rates are unlikely to rebound, as it is closely linked to future car ownership trends. Furthermore, millennials exhibit a threefold higher likelihood of relinquishing personal vehicle ownership in comparison to other generations.

In response to reduced car use, millennials have embraced alternative transportation modes, including carpools, public transit, and bicycling, at higher rates than older generations. However, it is worth noting that millennials also engage in less overall travel compared to previous generations at the same age, as reported by the National Household Travel Survey.

2.6.2 Adoption of Electric Vehicles

The United States holds the position of the third-largest Electric Vehicle (EV) market globally, with a notable 81% increase in sales recorded in 2018 (Nickkar, 2019). The total number of EVs in the country has now exceeded 1.5 million. However, despite the rapid growth witnessed in the EV sub-sector in recent years, EVs currently represent only 1.6% of private light vehicles in the United States, indicating their status as an emergent technology on the adoption curve. In the United States, EV ownership is predominantly observed among specific socio-economic groups, including males, middle-aged individuals, highly educated individuals, and high-income earners (Nickkar, 2019). These factors contribute to the demographic profile

Fig. 2.6 Electric Vehicle being charged in United States of America. (Source: Pixabay)

of EV owners in the country. The presence of a well-established charging infrastructure plays a crucial role in promoting the use of electric vehicles. The establishment of a widespread charging infrastructure that includes charging stations at homes, workplaces, and public areas serves to boost the confidence of electric vehicle (EV) drivers by extending the range of their vehicles. Additionally, it plays a crucial role in increasing the visibility of EV technology and raising public awareness. Studies on consumer behavior reveal that the availability of fast-charging stations and a well-developed charging market significantly influence the decision to purchase an EV. Notably, regions with a high adoption rate of EVs tend to have a minimum of 300 public charging points per million individuals (Bonges & Lusk, 2016).

While the United States has made significant strides in smart transportation development, it falls behind Germany in this regard. Nonetheless, plans are underway to improve the transportation landscape in the country with the goal of achieving sustainable transport planning (Fig. 2.6).

2.7 Transport Situation in Singapore

Singapore's urban transport system has gained global recognition for its efficient management and smooth traffic flow. The city-state is rapidly transforming its communities into smart cities through investments in innovative technologies, sustainable transport planning policies, and integration of transport modes (Singapore

Land Transport Authority [LTA], 2010). Originally a small fishing village, Singapore has experienced remarkable economic growth since its independence in 1965 and has become a major Asian metropolitan center (LTA, 2010). The country's limited land resources have posed challenges for planners and road users as congestion levels increased. To address this issue, the Area Licensing Scheme (ALS) was implemented in 1975. The ALS, the world's first toll system for road users during peak hours, resulted in a significant reduction in congestion levels by nearly 20% (Watson & Holland, 1978).

The success of the ALS was evident in various aspects. Firstly, it achieved the government's congestion reduction targets. Secondly, unlike many government initiatives, the ALS was financially profitable, generating revenues nine times the costs and providing a high rate of return on investment. Most importantly, Singaporean citizens accepted and supported the ALS as a necessary measure for improving the nation's transportation system (Watson & Holland, 1978). The Land Transport Authority (LTA), a statutory board under the Ministry of Transport, plays a crucial role in transportation planning and development in Singapore (LTA, 2009). The LTA focuses on expanding road networks to meet growing travel demand and utilizes Intelligent Transportation Systems (ITS) for traffic management (Wong, 2015). Public transport improvement is also prioritized, with the aim of making it the preferred mode of transportation (Palliyanni & Lee, 2017). To further enhance transportation, Singapore emphasizes the integration of land use and transport nodes to promote walking and cycling. Integrated Transport Hubs (ITH) with air-conditioned bus interchanges have been established, facilitating sustainable transport options. Transit-Oriented Development initiatives, such as the Jurong Lake District, exemplify the integration of mixed-use urban districts with comprehensive pedestrian and cycling networks (LTA, 2010). The district's development includes attractive commercial hubs interconnected by pedestrian walkways, public amenities, and public transport nodes. The LTA collaborates with the Urban Redevelopment Authority to create a comprehensive pedestrian and cycling network within the district, encouraging active mobility (LTA, 2009).

Overall, Singapore's commitment to smart transportation management, sustainable planning, and the integration of various transport modes has positioned it as a leading example of effective urban transport systems. The city-state's initiatives provide valuable lessons for other regions striving to address transportation challenges and achieve sustainable development.

2.8 Overview of Singapore Transport Policies Aimed at Sustainable Planning

Several measures aimed at improving the quality of transportation in terms of infrastructural development, smart technologies, and citizen participations have been adopted by Singapore stakeholders and government. Some of these include:

2.8.1 *Management of Transport Modes*

This can be categorized under the following headings:

2.8.1.1 Promotion of Public Transport

Based on a report from the Land Transport Authority (2010), public transportation in Singapore primarily relies on buses, taxis, and rail systems like MRT and LRT. Presently, there are two bus operators managing a fleet of 3394 buses operating across approximately 339 bus routes. The city also has seven private taxi companies with a combined fleet of 24,300 taxis. In 2010, public vehicles constituted a small portion of the overall vehicle population on the road, accounting for 16,300 buses and 26,000 taxis, which represented only 1.7% and 2.8% of the total vehicle population respectively. While the number of public vehicles has been growing, the rate of increase is lower compared to private cars (Land Transport Authority, 2010).

On another note, Singapore's Rapid Transit System (RTS) network includes Mass Rapid Transit (MRT) and Light Rail Transit (LRT). In 2010, the RTS network spanned 159 km, encompassing 4 MRT lines with 73 stations and 3 LRT lines with 33 stations. The density of Singapore's RTS in 2010 was 31 km per million individuals, similar to Hong Kong but slightly lower than Tokyo and New York, and nearly half of London. In terms of ridership, buses dominate public transport, accounting for 52% of the total daily trips in 2009. Out of a total of 5.84 million daily trips, buses recorded 3.06 million trips (Rahman & Chin, 2011, Fig. 2.7). In the same year, MRT, taxis, and LRT had daily average riderships of 1.83 million (31%), 0.86 million (15%), and 0.09 million (1.5%) respectively. However, when compared to road transport, RTS has experienced significantly higher growth rates between 2002 and 2009. The Singaporean government has implemented policies to increase the share of public transport during morning peaks from 59% in 2008 to 70% by 2020. Recent initiatives focused on enhancing bus services include peak hour bus lanes, full-day bus lanes, priority at signalized junctions, and mandatory give way to exiting buses from bus bays (Lim, 2010).

Overall, Singapore's public transport system is well-developed and characterized by technological innovation (Fig. 2.8).

2.8.1.2 Facilitation of Non-motorized Transport

As part of its transportation initiatives, Singapore is actively encouraging cycling and pedestrian walkability. The national government has strategically developed dedicated cycle routes and provided ample parking facilities to cater to both commuting and recreational needs (Rahman & Chin, 2011). To encourage the use of non-motorized transport, such as cycling, bicycle parking facilities have been installed at most MRT stations and bus interchanges. During off-peak periods, MRT

Fig. 2.7 Public transport with designated lane in Singapore. (Source: Piano luxury, 2023)

Fig. 2.8 Technological innovation in the public transport system in Singapore. (Source: Piano luxury, 2023)

trains and buses allow foldable bicycles on board, further enhancing their accessibility and usage by road users. Wong (2009) notes that Singapore has also developed dedicated cycling paths in seven towns by the end of 2014, reducing conflicts between motorists and pedestrians. Across the entire island, the Land Transport Authority (LTA) has established a robust system of well-planned footpaths, covered walkways, and an extensive network of overpasses and underpasses. These infrastructure components are designed to provide pedestrians with a comfortable and convenient walking experience, seamlessly connecting residential areas, commercial centers, educational institutions, and transportation hubs (Wong, 2009). The pedestrian overhead bridges are equipped with sheltered structures, and dynamic electronic displays are implemented to enhance pedestrian crossings.

2.8.1.3 Integration Among Passenger Modes

Efforts have been made to integrate rail and bus services to enhance transportation efficiency and convenience for commuters. This integration involves coordinating networks, facilities, fares, information, and schedules, with a focus on improving accessibility and interconnectivity. MRT and bus stations are strategically linked through well-designed walkways, allowing seamless transfers for commuters. In addition, selected MRT stations are equipped with taxi stands and bicycle parking facilities to accommodate different modes of transportation. To simplify the ticketing process, a universal fare card known as the 'EZ-link' is used for both train and bus travel. Furthermore, a distance-based through fare structure has been implemented to facilitate smoother transfers and eliminate fare penalties.

2.8.1.4 Promotion of Green Vehicles

Green vehicles, which utilize alternative fuels and propulsion technologies, have proven to emit lower pollution levels compared to conventional petrol and diesel vehicles (Land Transport Authority, 2010). These eco-friendly vehicles commonly operate on compressed natural gas (CNG), electricity, hydrogen, or hybrid systems that combine traditional fuels with green alternatives. Although the adoption of alternative fuels is still in its early stages in Singapore's road transport system, significant progress has been made. Currently, approximately 0.3% of buses and 1% of cars utilize alternative fuels such as petrol-electric and petrol-CNG, while about 10% of taxis run on bi-fuel CNG (Land Transport Authority, 2010). To further incentivize the use of green vehicles, Singapore introduced the Green Vehicle Rebate (GVR) scheme in 2001 (Land Transport Authority, 2009). This scheme offers registration fee offsets of around 40% for cars and 10% for motorcycles based on their open market values (Land Transport Authority, 2009). The GVR scheme aims to encourage individuals to choose environmentally friendly vehicles.

Additionally, Singapore has initiated the Electric Vehicle (EV) test-bedding program to assess infrastructure requirements, explore new business models related to

EVs, and identify opportunities for industry development (Land Transport Authority, 2010). This program aims to study the feasibility and potential of EVs in Singapore's transportation landscape. Through the implementation of schemes like the GVR and the exploration of EVs, Singapore is taking proactive measures to promote sustainable and environmentally conscious transportation options within the country (Land Transport Authority, 2009, 2010).

2.8.1.5 Promotion of Car Sharing Practices

Car sharing is widely recognized as a sustainable transportation solution that reduces the number of vehicles on the roads, resulting in reduced traffic congestion, pollution, and various social and economic impacts (Rahman et al., 2010). In the context of Singapore, a country known for its efficient public transport system, the implementation of effective car sharing practices can contribute to long-term sustainable urban transport (Tan, 2011). Currently, Singapore has three established car sharing companies, offering an alternative to car ownership, which is increasingly becoming financially burdensome for citizens. Users of car sharing services appreciate the convenience it offers without the typical challenges associated with owning a car, such as high costs, parking difficulties, insurance, and maintenance expenses. The availability of car sharing vehicles is concentrated in specific housing estates. Additionally, many organizations provide shared transportation services for their employees through the provision of company cars. Furthermore, certain high-traffic locations like shopping malls, commercial complexes, and schools have implemented their own shuttle services to facilitate shared transportation for their clients and students. These initiatives contribute to the overall availability and accessibility of shared transportation options in Singapore.

2.8.2 *Deployment of Smart Technology*

The achievement of sustainable development in the country's transportation system has been facilitated through the adoption of modern technologies. Singapore, being at the forefront of efforts to create smart cities, has implemented various strategies to accomplish this goal. These include:

2.8.2.1 Vehicle Emission Standard

Conventional petrol and diesel vehicles emit higher levels of pollution compared to green vehicles, which utilize alternative fuels and propulsion technologies (Land Transport Authority, 2010). These environmentally friendly vehicles operate on compressed natural gas (CNG), electricity, hydrogen, or hybrid systems that combine conventional and green fuels. While the adoption of alternative fuels is still in its early stages in Singapore's road transport system, significant progress has been made. Presently, approximately 0.3% of buses and 1% of cars run on alternative fuels such as petrol-electric and petrol-CNG, and around 10% of taxis operate on bi-fuel CNG (Land Transport Authority, 2010). To further encourage the use of green vehicles, Singapore introduced the Green Vehicle Rebate (GVR) scheme in 2001. This initiative provides incentives by offsetting approximately 40% of the registration fee for cars and 10% for motorcycles based on their open market values (Land Transport Authority, 2009). The GVR scheme aims to incentivize individuals to choose environmentally friendly vehicles.

Moreover, Singapore has launched the Electric Vehicle (EV) test-bedding program to assess infrastructure requirements, explore new business models associated with EVs, and identify opportunities for industry development (Land Transport Authority, 2010). This program aims to evaluate the feasibility and potential of EVs within Singapore's transportation landscape. Through initiatives like the GVR scheme and the EV test-bedding program, Singapore is actively promoting sustainable and eco-friendly transportation options, taking significant steps towards a greener future (Land Transport Authority, 2009, 2010).

2.8.2.2 Electronic Fare Collection

Smart fare collection systems play a crucial role in improving the efficiency and accuracy of transport transactions, as well as enhancing the overall travel experience for commuters. Singapore has implemented a sophisticated revenue collection system that incorporates smart technologies (Richmond, 2008). This system encompasses not only on-board and off-board fare payment for public transportation but also includes parking charge payments. To facilitate seamless and contactless payment of fares across various modes of public transport such as buses, MRT, and LRT, a tap-and-go fare card called 'EZ-link' has been introduced. Furthermore, an upgraded version of this smart card called 'Symphony for e-Payment (SeP)' has been developed, enabling users to make payments for other services including ERP, parking, and transactions at numerous retail outlets.

2.8.2.3 Electronic Road Pricing

According to a report by the Land Transport Authority (LTA) (2011), the control of vehicle usage is achieved through the implementation of Electronic Road Pricing (ERP). The ERP scheme utilizes a roadside communication system that interacts with on-board units installed in vehicles to deduct congestion charges accordingly. To participate in the scheme, each driver is required to possess a charging device equipped with a smart card, from which the charges are deducted. The ERP system offers flexibility in adjusting charges based on factors such as the time of day, location, vehicle type, and traffic conditions. As a result of these measures, the ERP scheme has successfully reduced traffic flows at the designated charging points. It is assumed that this reduction in traffic has subsequently led to a decrease in travel demand, potentially improving overall transport efficiency.

2.8.2.4 Smart Infrastructure Technologies

Singapore has embraced a range of intelligent technologies and systems to elevate its transportation infrastructure. One notable example is the Expressway Monitoring and Advisory System (EMAS), which functions as an astute incident management system on expressways (Sim, 2011). Additionally, the incorporation of intelligent road studs at major intersections has bolstered road safety measures. Singapore's MRT stations are also equipped with cutting-edge features like video surveillance, smart ticketing, and real-time train arrival displays. The city's traffic signal system operates seamlessly with smart technologies such as the Green Link Determining (GLIDE) System, transit signal priority for buses, Green Man Plus (GMP) technology, countdown timers, and audio signals for the disabled, all contributing to a more efficient transportation network. To enforce traffic regulations, Singapore has implemented speed cameras at key locations and red-light cameras at major intersections. Furthermore, the city utilizes approximately 280 advanced surveillance cameras known as J-Eyes at prominent signalized intersections, facilitating effective monitoring and enforcement of traffic rules and regulations.

Smart technologies have also been harnessed in Singapore's parking system. The Central Business District (CBD) boasts a smart parking guidance system, while housing estate multi-story car parks feature intelligent park gantry systems. Vehicle Parking Certificates play a pivotal role in managing parking arrangements across the city (Sim, 2011). Continuously striving for improvement, the Land Transport Authority (LTA) is presently conducting a trial of Closed-circuit Television (CCTV) systems to address street parking challenges. Collectively, Singapore's implementation of these intelligent technologies and systems underscores its unwavering dedication to enhancing transportation efficiency, fortifying road safety measures, and ensuring effective parking management throughout the city (Sim, 2011).

2.8.2.5 Smart Vehicle Technologies

Singapore's transportation system is enhanced by a comprehensive array of intelligent technologies and systems that collaborate to enhance efficiency and safety. The Expressway Monitoring and Advisory System (EMAS) serves as an intelligent incident management system, effectively handling traffic on expressways. Major intersections are equipped with intelligent road studs, while MRT stations feature video surveillance systems, smart ticketing systems, and electronic displays showcasing real-time train arrival information. The traffic signals are fully automated, incorporating advanced smart traffic signal technologies like the Green Link Determining (GLIDE) System, transit signal priority for buses (known as B-signal), Green Man

Plus (GMP) technology, countdown timers, and audio signals to assist the disabled. Singapore boasts speed and red-light cameras at significant intersections, along with approximately 280 advanced surveillance cameras known as J-Eyes at major signalized intersections. The parking system incorporates intelligent features such as a smart parking guidance system in the Central Business District (CBD), smart park gantry systems at multi-story car parks in housing estates, and Vehicle Parking Certificates. To combat street parking, the Land Transport Authority (LTA) is initiating a trial of Closed-circuit Television (CCTV) systems near potential roadsides (Sim, 2011).

2.8.2.6 Advanced Traveler Information

Online integrated traffic and transit map systems, such as Transit Link and OneMap, play a crucial role in providing travelers with alternative transport modes when faced with traffic congestion, accidents, and other road traffic issues. These systems assist users in planning and embarking on their trips. Additionally, SBS, SMRT, and LTA offer multi-modal journey planner portals that present travelers with the best routes and flexible choices. Real-time traffic information is readily accessible on the ONE.MOTORING portal, providing commuters with up-to-date traffic updates. Moreover, the public transport travel advisor offers pre-board information and bus arrival times can be obtained through the internet, SMS, and electronic display panels at selected bus stops and MRT station platforms. Ongoing efforts include the development of a smarter traffic prediction tool to enhance arrival time estimations. In the case of taxis, a smart system allows passengers to book rides using the internet, SMS, or phone. A convenient and centralized telephone number is available for all taxi booking facilities. On buses and RTS facilities, on-board displays provide real-time information on next stops and routes.

In conclusion, transportation systems in most developed countries prioritize sustainable planning. Major transport policies focus on encouraging the use of smart technologies by commuters and public transport systems. European countries and Singapore, in particular, advocate for non-motorized transport to reduce reliance on single-occupancy vehicles powered by fossil fuels, consequently minimizing harmful emissions. The integration of different transport modes is also emphasized to ensure commuters utilize multiple modes effectively. These measures contribute to the development of smart cities through the deployment of smart technologies, infrastructure, and policies that promote sustainable development.

References

Adriazola-Steil, C., & Schwedhelm, A. (2020). *Walking the talk: What can we learn from Germany's New Pedestrian Policy Framework?* https://thecityfix.com/blog/walking-talk-can-learn-germanys-new-pedestrian-policy-framework-2-claudia-adriazola-steil-alejandro-schwedhelm/. Accessed 10 Dec 2021.

Almenoar, M. (2010, June 30). Call for new ERP proposals. *Straight Times*.
Banister, D. (2005). *Unsustainable transportation*. Routledge.
Beatley, T. (2000). *Green urbanism: Learning from European cities*. Island Press.
Bicycle Dutch. (2013). *A little bit of Germany*. https://bicycledutch.wordpress.com/2013/10/03/a--little-bit-of-germany/. Accessed 18 Jan 2019.
BMVBS. (2000). *Urban development and urban policy in Germany*. German Federal Ministry of Transportation and Urban Development, Bundesamt fuer Bauwesen und Raumordnung.
Bonges, H. A., & Lusk, A. C. (2016). Addressing electric vehicle (EV) sales and range anxiety through parking layout, policy and regulation. *Transportation Research Part A: Policy and Practice, 83*, 63–73. https://doi.org/10.1016/j.tra.2015.09.011
Borade, S., Cohen, K., Davis, C., & Stockman, J. (2020). *Sustainability mobility in the United States: Challenges and opportunities from new technologies*.
Bormans, Y. (2018). *Statistical pocketbook 2018: Text. Mobility and Transport*. European Commission, August 6, 2018. https://ec.europa.eu/transport/factsfundings/statistics/pocketbook-2018_en
Bouton, S., Knupfer, S., Ivan M., & Swartz S. (2015). Urban mobility at a tipping point | McKinsey. McKinsey & Company, September 2015. https://www.mckinsey.com/business-functions/sustainability/our-insights/urban-mobilityat-a-tipping-point.
Buehler, R., & Pucher, J. (2009). Sustainable transport that works: Lessons from Germany. *World Transport Policy & Practice, 15*(1), 13–46.
Buehler, R., Pucher, J., & Kunert, U. (2009). *Making transportation sustainable: Insights from Germany*. Brookings Institution Metropolitan Policy Program.
Burchell, R., Lowenstein, G., Dolphin, W. R., & Gully, C. (2002). *Costs of sprawl 2000. TCRP Report 74*. Transportation Research Board. National Academy Press.
Divall, C., & Bond, W. (2003) Suburbanizing the masses: Public transport and urban development in historical perspective.
Downs, A. (2004). *Still stuck in traffic*. Brookings Institution Press.
Federal Highway Administration (FHWA). (2004). A review of pedestrian safety research in the United States and abroad. In F. H. Administration (Ed.), U.S. Department of Transportation.
Federal Highway Administration (FHWA). (2009) A resident's guide for creating safe and walkable. World Transport Policy & Practice. 15. 1. Communities. In F. H. Administration (ed.), U.S. Department of Transportation.
Fuerst, D., & Scholles, F. (2003). *Landes- und Regionalplanung Teil 1: Verwaltungswissenschaftliche Grundlagen*. Institut fuer Landesplanung und Raumforschung.
Hajdu, J. (1989). Pedestrian malls in West Germany: Perceptions of their role and stages in their development. *Journal of the American Planning Association, 54*, 325–335.
Hampel, C. (2021). German's new government aims for more sustainable transport. Electrive.com. Industry service for electric mobility. https://www.electrive.com/2021/11/25/germanys-new-government-aims-for-more-sustainable-transport. Accessed 8 Sept 2022.
Hass-Klau, C. (1993). *The pedestrian and city traffic*. Belhaven Press.
IBM. (2010). *IBM client reference: Singapore land transport authority*. http://www-07.ibm.com/innovation/my/exhibit/documents/pdf/1_Government_Singapore_Land_Authority.p. Accessed 18 Oct 2021.
Ingram, K. G., & Liu, Z. (1999). *Determinants of motorization and road provision*. The World Bank.
IRTAD. (2008). *Selected risk values for the year 2005*. International Traffic Safety Data and Analysis Group. OECD.
Kunzmann, K. (2001). State planning: A German success story? *International Planning Studies, 6*, 153–166.
Land Transport Authority (LTA). (2009). *Singapore land transport: Statistics in brief*.
Land Transport Authority (LTA). (2010). *Singapore land transport: Statistics in brief*.
Land Transport Authority (LTA). (2011, February 14). *Electronic road pricing*.
Lim, B. N. (2010). *Singapore's land transport system: Enabling the growth of a nation, world urban transport leaders' summit*.

Low, N., & Gleeson, B. (2002). *Making urban transport sustainable*. Palgrave Macmillan.

Martin, E., Shaheen, S., Zohdy, I., Chan, N., Bansal, A., Bhattacharyya, A., Aly, T., Yelchuru, B., Finson, R., & Yeung Yam Wah, C. (2016). *Understanding travel behavior: Research scan*. U.S. Department of Transportation, Federal Highway Administration (FHWA).

Newman, P., & Kenworthy, J. (1999). *Sustainability and cities*. Island Press.

Newman, P., Kenworthy, J., & Laube, F. (1999). *An international sourcebook of automobile dependence in cities, 1960–1990*. University Press of Colorado.

Newman, P., Beatley, T., & Boyer, H. (2009). *Resilient cities. Responding to peak oil and climate change*. Island Press.

Nickkar, A. (2019). *Electric vehicle ownership factors and commuting behavior in the United States*.

Nivola, P. S. (1999). *Laws of the landscape. How policies shape cities in Europe and America*. Brookings Institution Press.

Palliyanni, S., & Lee, D. H. (2017). Sustainable transport policy – An evaluation of Singapore's past, present and future. *Journal of Infrastructure, Policy and Development, 1*(1), 112–128. https://doi.org/10.24294/jipd.v1i1.23

Piano luxury. (2023). *Useful! How to Ride the Bus in Singapore and Route Map*. https://pianoto-hikouki.com/en/singapore-bus. Accessed 20 Feb 2023. https://pixabay.com/photos/amsterdam-smartcar-electric-car-eco-916561/. Accessed 20 Feb 2023.

Polzin, S. (2018). Changing travel behavior: We are traveling less, and more. *Planetizen – Urban Planning News, Jobs, and Education*. https://www.planetizen.com/blogs/100166-changing-travel-behavior-we-are-travelingless-and-more

Pucher, J. (1988). Urban travel behavior as outcome of public policy. *APA Journal, Autumn, 1988*, 509–520.

Pucher, J. (1995). Urban passenger transport in the United States and Europe: A comparative analysis of public policies. Part 2. Public transport, overall comparisons and recommendations. *Transport Reviews, 15*, 211–227.

Pucher, J., & Buehler, R. (2008). Making cycling irresistible: Lessons from The Netherlands, Denmark, and Germany. *Transport Reviews, 28*.

Pucher, J., & Dijkstra, L. (2003). Promoting safe walking and cycling to improve public health: Lessons from The Netherlands and Germany. *American Journal of Public Health, 93*.

Pucher, J., & Kurth, C. (1996). Verkehrsverbund: The success of regional public transport in Germany, Austria and Switzerland. *Transport Policy, 2*.

Rahman, M. H., & Chin, H. C. (2011). Sustainable urban transport in Singapore: A balanced scorecard. *OIDA International Journal of Sustainable Development, 2*(10), 20–42. Available at http://www.ssrn.com/link/OIDA-Intl-Journal-Sustainable-Dev.html

Rahman, M. H., Chin, H. C., & Haque, M. M. (2010). Sustainability in road transport: An integrated life cycle analysis for estimating emissions. In *Proceedings: International conference on sustainable built environment*.

Richmond, J. E. D. (2008). Transporting Singapore: The air-conditioned nation. *Transport Reviews, 28*(3), 357–390.

Rodrigue, J. P., Comtois, C., & Slack, B. (2012). *The geography of transport systems* (3rd ed.). Published by Routledge.

Schafer, A., & Victor, D. G. (2000). The future mobility of the world population. *Transportation Research Part A, 34*, 171–205.

Schmidt, S., & Buehler, R. (2007). The planning process in the US and Germany: A comparative analysis. *International Planning Studies, 12*, 55–75.

Sharma, S. N. (2022). *What is mixed land use?* https://www.quora.com/What-is-mixed-land-use. Accessed 15 Dec 2022.

Shyshkina, T. (2020). New York City, USA. https://unsplash.com/photos/-JyTddA7vec. Accessed 15 Feb 2021.

Sim, R. (2011, September 20). LTA to run trial on using CCTV cameras to curb illegal parking. *Straight Times*.

Sung, J. H. K. (2008). Environmentally sustainable transportation: Singapore's approach. *Tech Monitor*, 30–34.

Tan, C. (2011, July 20). Car-sharing back in the fast lane. *Straight Times*.

The Business Times. (2010, September 14). *Singapore sounds out oil firms on ultra-low sulphur petrol, diesel.*

Tolley. (2003). *Sustainable transport: Planning for walking and cycling in urban environments*. Woodhead Publishing.

Topp, H. (1993). Parking policies to reduce car traffic in German cities. *Transport Reviews, 13*, 83–95.

VDV. (2005). *VDV Statistik 2005*. Verband Deutscher Verkehrsunternehmen.

Verron, H., & Friedrich, A. (2004). Environmentally sustainable transport in Germany. *European Journal of Transport and Infrastructure Research, 4*(1), 71–98.

Vuchic, V. (1999). *Transportation for liveable cities*. Center for Urban Policy Research (CUPR).

Vuchic, V. (2005). *Urban transit: Operations, planning, and economics*. Wiley.

Watson, P. L., & Holland, E. (1978). *Relieving traffic congestion: The Singapore area license scheme*. World Bank.

Wendt, C. (2018). Germany's roadblock: Infrastructure investment long overdue. https://www.walleniuswilhelmsen.com/insights/germanys-roadblock-infrastructure-investment-long-overdue. Accessed 12 Mar 2023.

Wong, K. (2009). Environmentally sustainable transport – Singapore's experience, fourth regional EST Forum, Seoul, Korea, 24–26 February, 2009.

Wong, W. L. (2015). *Comparison and review on bus priority schemes between Hong Kong, Singapore and London*. HKU Theses Online (HKUTO). The University of Hong Kong, Pokfulam.

Chapter 3
Transport Situation in the Global South: Insights from Nigeria, South Africa and India

Oluwayemi-Oniya Aderibigbe, Samson Olawale Fadare, and Trynos Gumbo

3.1 Introduction

The role of transport in the economies of African nations cannot be overemphasized, facilitating the movement of people and goods across different regions. Understanding the evolution of transportation in Africa requires a historical perspective, which can be categorized into three distinct eras: the precolonial period, the colonial period, and the post-colonial period. During the precolonial period, infrastructure development was limited, resulting in a lack of well-established transport networks. The colonial period witnessed the emergence of distinct areas, with cities serving as administrative centers for rulers and suburban and rural regions primarily inhabited by indigenous populations. It was during this period that significant investments were made in transport infrastructure. In many countries, the urban master plans still adhere to the layouts established decades ago by colonial powers. Road construction primarily focused on connecting areas where civil servants resided, with limited emphasis on linking residential areas and agricultural lands. The exploitation of natural resources played a prominent role in the colonization of Africa. As a result, laborers and transportation infrastructure were required to transport valuable commodities such as cotton, rubber, diamonds, and gold. Railways were constructed to facilitate the transportation of these resources from inland mines and farms to coastal areas, where they could be exported to Europe.

3.2 Transportation Development in Nigeria

Regardless of a nation's industrial capacity, population size, or technological development, transportation remains an essential necessity for every country. In Nigeria, there are three primary modes of transportation (land, water, and air) that serve the

O.-O. Aderibigbe et al., *Emerging Technologies for Smart Cities*,
https://doi.org/10.1007/978-3-031-66943-9_3

country's economic, social, and environmental development needs. These modes include road, water, rail, and air transport. The development trends of these various forms of transportation can be summarized using key statistics. In 1925, there were approximately 1300 km of motorable roads, which increased to 40,000 km in 1950 and further to 100,000 km in 1975. During the same period, the number of newly registered motor vehicles also rose, from 3130 in 1930 to 6240 in 1950, 33,290 in 1965, and 75,215 in 1974. By 1980, Nigeria had a total of 114,768 km of roads, and between 2003 and 2006, the Federal Government added an additional 34,340.95 km of roads. Furthermore, data from uniform licensing scheme production statistics for the years 2002 to 2006 indicate the following vehicle registration figures: 692,987 in 2002, 565,582 in 2003, 480,251 in 2004, 506,162 in 2005, and 762,016 in 2006 (NBS, 2008). These figures demonstrate an increase in the number of passengers and infrastructure for road transportation.

However, while road transportation has experienced growth, rail transportation has witnessed a decline. The railway system reached its peak in 1960/61, carrying 7.9 million passengers, but the number declined to 4.7 million passengers in 1973. Anyanwu et al. (1997) presented passenger traffic data as follows: 9.6 million in 1960–1971, 6.1 million in 1972–1978, 8.2 million in 1979–1982, 11.5 million in 1983–1987, and 4.4 million in 1988–1992, indicating a peak between 1983–1987. In the first decade of the twenty-first century, the figures were 987,088 in 2002, 1,622,271 in 2003, 1,751,159 in 2004, and 752,482 in 2005 (NBS, 2008). This implied that even though there has been growth in the road construction while there has been a decline in the rail development in Nigeria.

3.3 Role of Transportation in the Development of Nigeria Economy

Transportation plays a vital role in the economic development of nations, and this holds true for Nigeria as well (Glen & Reno, 2009). The uneven distribution of natural resources and essential services across the country necessitates the transportation of these resources to areas where they are in demand but not readily available. This transportation requirement forms the basis for economic growth and activities within the nation. The significance of transportation to the Nigerian economy can be observed through various aspects, as highlighted by Glen and Reno (2009). These include:

- Cost savings for individuals who use public transportation or switch from automobiles, resulting in shifts in consumer spending.
- Reduced traffic congestion for car and truck users, leading to direct travel cost savings in terms of time, businesses, and households. Reduced congestion also contributes to cost savings for businesses through improved worker productivity and reliability.

- Increased business productivity resulting from access to a wider labor market with diverse skills, facilitated by reduced traffic congestion and expanded transit services.
- Indirect impacts of business growth on supplies and induced impacts on spending of worker wages, which lead to additional regional business growth.
- At a national level, cost savings and increased productivity have the potential to enhance competitiveness in international markets.

Overall, transportation plays a multifaceted role in the Nigerian economy, encompassing cost savings, enhanced productivity, and improved regional and international competitiveness.

3.4 Urban Road Transportation Situation in Nigeria

Transportation systems encompass various modes such as roads, railways, water transport, and air transport. The road system comprises transit systems and paratransit services, including private automobiles and pedestrians. According to Rodrigue, Comtois, and Slack (2013), the rapid urban development witnessed globally has resulted in an increased number of travelers and cargo within cities, consequently intensifying mobility demands. This population growth has exacerbated transportation challenges in both developed and developing countries (Fawcett, 2000). In Nigeria, the road system is categorized into three major types: Federal Trunk A Roads, Regional/State Trunk B Roads, and Provincial/Local Government Trunk C Roads. Under this classification, separate responsibilities for road planning, construction, and maintenance are assigned to the Federal, Regional/State, and Provincial/Local Governments, respectively. Unfortunately, this allocation system lacks equity and remains in effect today. Consequently, Local Governments with limited resources and personnel bear the largest burden of road construction and maintenance, while the federal government, with greater manpower and financial capacity, assumes minimal responsibility. As a result, road conditions in many parts of the country, particularly rural areas, are deplorable.

The road transport infrastructure in Nigeria encompasses various modes of transportation, including buses, cars, trailers, walking, bicycles, and even animals and pedestrians. However, these roads suffer from traffic congestion issues (see Fig. 3.1) and lack pedestrian walkways. This inadequacy renders active modes of transportation, such as walking and cycling, unsafe for pedestrians. Moreover, a significant portion of Nigeria lacks adequate road coverage, and the existing roads are characterized by poor quality and maintenance. They are often untarred, narrow, winding, and seasonally unusable, especially during the rainy season when agricultural activities in rural areas peak (Aderibigbe & Gumbo, 2022a, b).

Rapid urbanization and increasing poverty pose significant challenges for developing countries in the new millennium (Salau, 1992; Mabogunje, 2002). This is particularly evident in Nigeria, where there has been a continuous influx of

Fig. 3.1 Road traffic congestion in Lagos Nigeria. (Source: United Nations Report on Non-Motorised Transport in Nigeria)

population into urban areas for over 50 years without corresponding infrastructural development (Salau, 1992; Mabogunje, 2002). Consequently, the economies of developing countries have experienced stagnant and unrewarding growth (Salau, 1992; Mabogunje, 2002). Urban development and infrastructure have been neglected in most Nigerian cities, leading to overburdened public services and inadequate social infrastructure (Filani, 2003; Oyesiku, 2002; Odufuwa, 2003; Aderibigbe, 2023). In recent years, transportation infrastructure has been deprioritized in favor of poverty reduction programs, resulting in limited investment in transport maintenance and the construction of new infrastructure (Filani, 2003; Oyesiku, 2002; Odufuwa, 2003; Aderibigbe, 2023).

As a result of this population explosion, transport provision has become inadequate, leading to deteriorating roads and substandard vehicles (Filani, 2003; Oyesiku, 2002; Odufuwa, 2003; Aderibigbe, 2023). Public transportation in Nigerian cities often suffers from overcrowding, poor services, and an unhealthy travel environment (Filani, 2003; Oyesiku, 2002; Odufuwa, 2003; Aderibigbe, 2023). The transportation situation in many Nigerian cities has reached a crisis point due to years of neglect by successive administrations (Filani, 2003; Oyesiku, 2002; Odufuwa, 2003; Aderibigbe, 2023). A significant proportion of households in these cities have incomes below the poverty line, affecting their ability to procure new vehicles and leading to a decline in mobility provision (Osinubi, 2003; Oyesiku, 2004; Odufuwa, 2003; Ademiluyi and Gbadamosi, 2004). Road maintenance in Nigeria has also been a concern, with a substantial amount of investment dissipated due to a lack of maintenance (Heggie, 1995). For instance, a World Bank investigation revealed that by 1990, nearly a third of the $150 billion invested in roads in Nigeria had been wasted (Heggie, 1995). Although the World Bank approved a credit for road construction and maintenance in Nigeria, the country continues to face significant transport challenges such as traffic congestion and accidents (Dosunmu & Mogbojuri, 2019).

According to a report by the Federal Road Safety Cooperation, FRSC (2011), depicted in Fig. 3.2, it is evident that road transport infrastructure in Nigeria has received inadequate investment. The report highlights a pressing need for substantial infrastructure upgrades, as a significant proportion of roads in the country remain unpaved. Moreover, Nigerian cities are grappling with a growing scarcity of urban transport infrastructure and services, limiting access for an increasingly smaller portion of the population. According to the data presented in Table 3.1, it is evident that a significant portion of Nigeria's national road networks, comprising less than 50%, are in a deteriorated state mainly because of insufficient maintenance efforts. Despite having approximately 113,000 km of surfaced roads, Nigeria faces challenges in maintaining them adequately, which has been identified as a contributing factor to the country's high rate of traffic fatalities. In 2004, the Federal Roads Maintenance Agency (FERMA) in Nigeria initiated patching activities to address the condition of the 32,000-km Federal roads network. However, these efforts face difficulties during the rainy season and due to the scarcity of road construction and maintenance equipment. The substandard condition of Nigerian roads imposes

Fig. 3.2 Typical overview of road conditions in Nigeria

Table 3.1 Road conditions in some selected Nigerian cities

Cities	% Unpaved roads	Maintenance cost/vehicle per month (USD)	Paved road conditions		
			Good	Fair	Bad
Abeokuta	62.5	51	32.1	32.5	35.4
Ijebu-Ode	51.3	46	31.5	47.5	21.0
Lokoja	54.6	35	48.3	34.5	17.2
Umuahia	58.6	48	39.4	40.2	20.4
Uyo	50.2	40	38.6	39.5	21.9
Zaria	51.4	52	38.2	34.3	27.5
Ibadan	69.5	49	30.2	43.5	26.3
Lagos	63.6	52	50.9	19.3	29.8
Kaduna	57.4	53	53.1	39.7	7.2

Source: Oyesiku (2002)

additional vehicle operating costs of N53 billion on road users, as reported by Road Vision in 2000. Urgent attention and increased investment are required to address these issues and enhance the state of road transport infrastructure in Nigeria.

3.5 Road Transportation Infrastructure in Nigeria

Efficient infrastructure facilities play a pivotal role in national development, acting as catalysts for urbanization. Urban transport systems rely heavily on robust infrastructures, encompassing roads, intra-city rail lines, ferry terminals, and complementary facilities like bus stops, service garages, and road signs, along with provisions for non-motorized transport. While urban transport predominantly revolves around road-based systems, some Nigerian cities also feature prominent rail and water infrastructures. However, urban road infrastructures often face challenges such as car-centric planning and design, poor physical conditions, inadequate management, limited connectivity, rampant misuse, and low productivity. In Nigeria, government ownership and management are prevalent in urban transport infrastructures, but existing funding mechanisms have proven inadequate, leading to infrastructure deterioration and difficulties in expanding provisions. Unfortunately, urban centers in Nigeria generally overlook the importance of urban transport infrastructure, including non-motorized modes like walking and bicycling. Interchange points, bus stops, terminals, parking facilities, and other essential elements are either lacking, insufficient, or poorly located (Federal Republic of Nigeria, 1995, Figs. 3.3 and 3.4). In most cases, the availability of quality infrastructure attracts investors to a nation, and as such, most developed countries invest in good infrastructure so as to bring about positive development to their country. Infrastructure has been a persistent concern in Nigeria, a developing nation, as it falls behind several African countries in terms of both quality and coverage. Particularly, rural

Fig. 3.3 Risky open drain along major road. (Source: United Nations Report on Non-Motorised Transport in Nigeria)

Fig. 3.4 Lack of pedestrian walkway. (Source: United Nations Report on Non-Motorised Transport in Nigeria)

areas, where the majority of the population resides, suffer from a significant lack of basic transportation infrastructure. Nigeria's major road network includes 32,000 km of Federal highways, seven major bridges spanning the Niger and Benue Rivers, the Lagos ring road, the third mainland axial bridge, 30,500 km of state roads, and 130,000 km of local roads (Buhari, 2000). According to Buhari (2000), as of June 1996, only 50% of the Federal roads and 20% of the State roads were in acceptable condition, and merely an estimated 5% of the local rural roads were suitable for motorized transport.

In present-day Nigeria, a significant portion of the population resides in areas lacking proper road infrastructure, with the majority of roads being in a dilapidated state (refer to Fig. 3.2). This unfortunate condition has led to increased production and transportation costs, hindering the competitiveness of the Nigerian economy both domestically and internationally. The consequences of inadequate equipment and facility maintenance are evident across all sectors. Roads are in unsatisfactory conditions, necessitating reconstruction, while insufficient vehicle replacement and maintenance contribute to high social costs in terms of pollution and operating expenses. Consequently, these escalating operating costs and diminishing net revenues make timely vehicle replacement challenging. The railway system has also suffered severe deterioration due to insufficient resources allocated for track, rolling stock, and maintenance facilities. Similar problems afflict inland waterways, diminishing their functionality. According to Olomola (2003), the insufficient provision of transport infrastructure and services is a key factor explaining poverty levels in various Nigerian communities, whether urban or rural. Transport-related issues such as poor road conditions, fuel problems (high prices, shortages), traffic congestion, inadequate mass transit options, high maintenance costs, and aged vehicles exacerbate the situation. Inadequate transport facilities and restricted accessibility to essential amenities like markets, hospitals, and water sources have dire implications for poverty. Urgent policy measures are necessary to address these pervasive transport problems. Road transport remains the primary mode of transportation in Nigeria, handling over 90% of goods and passenger movements (Filani, 2002). Although the Rolling Plans of 1991–1993 and 1994–1995 saw a slight shift toward investment in water and air transport, the road sub-sector still accounted for the majority of total investment.

In Nigeria, both major urban centers and rural areas suffer from inadequate road conditions, exacerbating the misuse of existing urban transport infrastructures on a wide scale. A notable issue is the encroachment of traders onto major roads, leading to traffic congestion as drivers contend with them for road space. Furthermore, bus stops, where they exist, are occupied by the homeless, access roads are used for improper waste disposal, and various transportation elements such as railings, highway barriers, and traffic lights are subject to rampant vandalism. A functional transportation system should ideally provide reliable, safe, comfortable, and easily accessible services at a reasonable cost to meet the temporal and spatial travel demands of society. To achieve these essential service attributes, adequate infrastructure becomes imperative.

According to Connor (1993), a city's transport infrastructure needs encompass various aspects:

- Bus Transport: This category encompasses various components essential for bus operations, including buses themselves, maintenance and emergency vehicles, terminals and depots, garages and workshops, office buildings, training facilities, road enhancements to prioritize buses, bus stops and shelters, fare collection equipment, as well as software for routing, inventory management, and data analysis.
- Rail Transport: The requirements for rail transport encompass a range of elements such as tracks, stations (including fare collection equipment, escalators, and elevators), rolling stock (passenger cars, work trains, rider cars, hopper cars), tunnel lighting, signaling and communication systems, power equipment substation and circuit breakers, car maintenance and overhaul yards, depots, security systems, and safety systems (fan plants, pump rooms, and fire suppression systems). However, in Nigeria, urban road infrastructure planning and designs have predominantly focused on meeting the demands of private cars, often overlooking the needs of buses and non-motorized modes. This has resulted in inadequate maintenance of urban roads, with approximately 60% of them being in poor condition. Consequently, this situation has led to escalating traffic congestion, increased vehicle operating costs, and the unviability of public transport (National Transport Policy for Nigeria, 2003).

3.6 Multi Modal Transport System in Nigeria

Multimodal transport plays a crucial role in establishing a more efficient and sustainable transportation system by diversifying the mobility of passengers and goods, reducing reliance on roads, and optimizing infrastructure utilization at reduced costs (Joshi & Maclean, 1995). Achieving cost-effective transport and logistics requires an optimal combination of various transportation modes, including active travel (walking/cycling), public transport, maritime, and rail transport (Jain & Geetam, 2020). The history of rail transport in Nigeria dates back to 1898 when it was initially established as a government department under the management of the civil service. However, the reliance on government support and direction hindered its development. In 1955, the Nigerian Railway Corporation was established to take over the administration of the railway department (Ali, 2013). Despite some investments made by different regimes, such as the Shagari regime's investment of $32 million in 1983 and the subsequent investments by Babangida, Abacha, and Obasanjo (Oye et al., 2016), rail transport in Nigeria has not witnessed substantial growth in comparison to road transport. The lack of significant progress in achieving sustainable development in the rail and maritime transportation sectors remains a challenge (Oye et al., 2016). According to data released by the National Bureau of Statistics in 2012, the majority of cargo landings in Nigeria occur at the Lagos ports, namely the Apapa Port and Tin Can Island Port (TCIP). In the years 2008–2010, the combined quantity of cargo received at all other ports pales in comparison to the cargo volume handled by these two ports (see Figs. 3.5 and 3.6). Consequently, excessive pressure on the facilities of the Lagos ports has resulted in poor service delivery, delays, and inadequate support for the development of a flourishing multimodal transport system (National Bureau of Statistics, 2012).

Fig. 3.5 Maritime transport in Nigeria

Fig. 3.6 An overview of congested maritime transport in Apapa Port Nigeria

3.7 Integrated Transport System Policies In Nigeria: The Case of Non-motorized and Public Transport Use in Nigeria

According to the United Nations Report (United Nations, 2023), the urban population of the Federal Republic of Nigeria is estimated to be 100 million. The rapid increase in population coupled with the absence of non-motorized transport (NMT) infrastructure has resulted in severe congestion due to the heavy reliance on personal motor vehicles. Therefore, there is an urgent need to develop well-designed transport systems that prioritize safe and efficient mobility for all urban residents. Non-Motorized Transport (NMT) plays a crucial role in achieving a more efficient and environmentally friendly transportation system. NMT offers a cost-effective, rapid, accessible, convenient, and safe mode of transport (Brussel, 2014). Several measures can be implemented to promote NMT, including the construction of sidewalks and cycle paths, the provision of dedicated lanes for cyclists and pedestrians, the integration of NMT networks with employment and commercial centers, and the establishment of continuous routes equipped with signals and safety barriers. Creating attractive and inviting environments that encourage walking is essential as people are more likely to engage in walking activities in clean and pollution-free surroundings (Belwal, 2016).

In Nigeria, Lagos stands out as one of the fastest growing cities globally, with an estimated population ranging from 18 to 20 million and an annual growth rate of 6% (United Nations Report, 2023). The transportation system in Lagos heavily relies on road infrastructure, which is currently overwhelmed. Many streets and roads lack proper footpaths, and pedestrian paths are often encroached upon by parked vehicles (see Fig. 3.7). Informal paratransit modes such as danfos and taxi services like kekes (three-wheelers) and okadas (motorcycle taxis) are widely utilized by the majority of residents in Nigeria. Due to the absence of dedicated pedestrian spaces, pedestrians are forced to share the road with fast-moving vehicles, resulting in conflicts between pedestrians and motorists and an increased risk of road traffic accidents. The combination of high vehicle speeds and the absence of dedicated spaces poses a significant safety concern for NMT users. As vehicle speeds exceed 30 km/h, the probability of NMT fatalities rises significantly. Consequently, NMT users are the most vulnerable and exposed group in terms of road safety. In Nigeria, various policy initiatives have been implemented to promote non-motorized transport (NMT) and these include:

- Street design standards: The government has introduced street design standards aimed at developing "complete streets" (Fig. 3.8). These streets prioritize walking and cycling by providing dedicated spaces for non-motorized transport modes. Additionally, designated areas for street vending and meeting points will be created, while free parking for cycles and on-street parking for personal motorized vehicles will be provided. To enhance pedestrian safety and prioritize

Fig. 3.7 Lack of parking facilities at major bus stops. (Source: United Nations Report on Non-Motorised Transport in Nigeria)

Fig. 3.8 Government approach towards supporting Non-Motorized Transport (NMT) in Lagos Nigeria. (Source: United Nations Report on Non-Motorised Transport in Nigeria)

public transport, traffic calming measures and at-grade crossing facilities will be implemented. The inclusion of street trees will offer shade for pedestrians and cyclists, and modern urban designs with appropriate street furniture will contribute to attractive public spaces.

- Land use-transport integration: The government intends to adopt land use regulations and policies that encourage higher density commercial and residential development within walking distance of major public transport corridors. This integration of land use and transport aims to promote non-motorized transport by directing urban growth along public transport routes and reinforcing existing urban centers. The objective is to provide convenient access to good public transport options.
- Efficient public transport systems: Efficient and high-quality public transport systems are crucial in promoting NMT. The government is committed to supporting the introduction of affordable, reliable, convenient, safe, and accessible public transport systems for all users, regardless of age, income, gender, ability, or social standing. The responsibility for ensuring a high-quality public transport system lies with the government, whether operated directly or by private service providers. Additionally, regulated paratransit services play a significant role in providing informal and flexible public transport options to urban residents and visitors. By regulating routes, fares, vehicle design, and technology, paratransit services can improve passenger convenience, safety, and complement formal public transport networks.
- Regulation and enforcement: The government recognizes the importance of regulation and enforcement in ensuring the success of shared mobility services. To this end, the government aims to establish registered associations for all keke operators and implement an identification system for them. Collaboration with these associations will enable expanded access to safety and customer service training for operators. Monitoring operations will encourage helmet usage, safe driving practices, and the establishment of designated stands that maintain clear space on footpaths and cycle tracks.

3.8 Transportation Situation in Africa: Experience from South Africa

South Africa has made significant progress in developing an effective transportation planning system, with a focus on innovative public transport initiatives. These initiatives, including the Bus Rapid Transit (BRT) and High-Speed Rail, have been implemented to enhance the transportation system and improve mobility for residents (Smith et al., 2017). The country has also prioritized the promotion of non-motorized transport, commonly referred to as active travel. This is evident in the establishment of numerous cycle lanes and pedestrian sidewalks, which serve as crucial links between pedestrians and other modes of transportation (Johnson & Johnson, 2019). In its pursuit of world-class cities, South Africa has adopted strategies aligned with international standards in transportation planning (Marais, 2020). Notably, the Rea Vaya BRT system and Gautrain high-speed rail have been successfully implemented across three metropolitan municipalities in the Gauteng province (Smith et al., 2017). These projects have significantly improved public transportation options and connectivity within the region. Additionally, the City of Johannesburg has invested in pedestrian transport infrastructure, including the development of cycling lanes and sidewalks, to enhance accessibility and safety for pedestrians (Marais, 2020). Overall, South Africa's transportation planning initiatives, encompassing the BRT system, high-speed rail, and investments in non-motorized transport infrastructure, demonstrate the country's commitment to achieving sustainable and efficient urban mobility (Johnson & Johnson, 2019). These efforts contribute to the continuous development of world-class cities within South Africa, benefiting both residents and visitors alike.

3.9 The State of Transport Infrastructure in South Africa

The development of road networks and other transport infrastructure is crucial for improving transportation systems within a country. In South Africa, the Department of Transport, in collaboration with local government, is currently working on the formulation of the Public Transport Infrastructure Fund to enhance integrated planning in land use and transport (Department of Transport, 2008). This initiative includes the implementation of Integrated Transport Plans and effective strategies, as well as the identification of strategic corridors that align with growing passenger demands. Consequently, there will be a need for periodic upgrades and improvements in road infrastructure. One of the existing challenges in South Africa's transportation system is the absence or limited availability of park-and-stop sites near business centers. Furthermore, the few park and ride facilities that do exist lack proper security measures and necessary amenities for promoting transportation development in the country (Department of Transport, 2008). Another issue affecting transportation efficiency and safety is the inadequacy of signage systems in addressing traffic management warnings. While pedestrians often associate pedestrian facilities solely with sidewalks, these facilities encompass a broader range of services and infrastructure. Pedestrian facilities include traffic control devices, curb ramps, grade separations (such as overpasses and underpasses), crosswalks, and design features intended to encourage pedestrian travel. These facilities are designed to facilitate the mobility of both pedestrians and motorists and should be incorporated into the public right-of-way (Department of Transport, 2008).

Efficient transport infrastructure provision in South Africa is hindered by outdated road designs from the 1960s. These designs featured guardrails that separated roads to prevent pedestrian crossings and narrow pavements. In addition, unwelcoming gyratory systems and dual carriageways limited pedestrian freedom on the road (Department of Transport, 2008). To address these issues, new traffic planning and road engineering approaches should consider pedestrians as integral members of the road community, while land use patterns should encourage the use of public transport in both urban and rural areas. Road networking is a critical component of transport infrastructure, with South Africa currently having approximately 12 million motor vehicles, averaging 16 vehicles per km (Department of Transport, 2008). The provincial road network spans approximately 222,951 km, consisting of 170,837 km of unpaved roads and 52,114 km of paved roads. Gravel roads account for 77% of the provincial road network. Table 3.1 illustrates that the average traffic density on the provincial road network, excluding national and municipal roads, is 55 vehicles per km. Gauteng and the Western Cape experience the highest traffic densities, leading to increased maintenance requirements. Conversely, the Northern Cape and Free State have lower traffic densities, at 10 and 14 vehicles per km, respectively. The Eastern Cape and Free State possess the longest road networks, each spanning over 40,000 km.

The comparison of paved and unpaved roads across South African provinces, as shown in Table 3.2, reveals that the number of paved roads is lower than unpaved roads in all provinces except Gauteng. This observation is supported by the World Economic Forum's Global Competitiveness Report 2018–19, which ranked the quality of South Africa's road network as 47th out of 141 surveyed countries (World Economic Forum, 2018–19). This ranking represents an improvement of three places compared to the previous year's report, where South Africa was ranked 58th (World Economic Forum, 2017–18). To maintain competitiveness, it is crucial for the country to invest in the construction and maintenance of road infrastructure

Table 3.2 Road conditions across the provinces in South Africa

Province	Paved extent	Unpaved extent	Total provincial road network	Total number of vehicles[a]	% Share of vehicles	Road densities
Eastern Cape	3808	36,345	40,153	816,898	6.6%	20
Free State	6707	39,149	45,519	628,885	5.1%	14
Gauteng	4284	1356	5640	4,737,590	38.5%	840
Kwazulu-Natal	8084	23,254	31,338	1,638,774	13.3%	52
Limpopo	7583	13,473	21,056	698,340	5.7%	33
Mpumalanga	5453	8373	13,826	882,585	7.2%	64
Northern Cape	4987	23,963	28,950	278,182	2.3%	10
North West	4929	14,550	19,479	620,393	5.0%	32
Western Cape	6616	10,374	16,990	1,991,566	16.2%	117
Total	52,114	170,837	222,951	12,293,213	100.0%	55

Source: National and provincial Departments of Transport (Road Asset Management System)
[a]Vehicles per kilometre

(World Economic Forum, 2018–19). However, it is equally important to develop the capacity required to efficiently and effectively plan and allocate resources for these projects. The South African Institute of Civil Engineers (SAICE) Infrastructure Report Card 2017 highlighted a slight improvement in the paved provincial road network, albeit with a low rating of "D" and still at risk of failure (SAICE, 2017). On the other hand, the unpaved provincial roads received an "E" rating, rendering them unfit for their intended purpose (SAICE, 2017). The decline in the overall condition of provincial road networks can be attributed to the loss of experienced road professionals over the past two decades, as noted in the SAICE report (SAICE, 2017). This deterioration not only hampers the state of transport infrastructure but also compromises road safety. Therefore, it is imperative to prioritize the improvement of transport infrastructure, particularly the road systems, and ensure the provision of adequate support for non-motorized transport in both urban and rural areas of the country.

3.10 Non-motorized and Public Transport Situation in South Africa

Transportation plays a crucial role in facilitating development by providing access to essential resources and opportunities such as goods, services, education, employment, and social interactions (World Bank, 2017). Inadequate transportation systems in rural areas of Africa, where non-motorized transport is predominant, have led to challenges like limited market integration, inadequate education and healthcare services, low productivity, and sluggish economic activity at the regional and local levels (Tumwebaze, 2017; Musakwa & Gumbo, 2017a, b). Effective transportation governance requires collaboration among all three levels of government (local, regional, and national) (Marshall & Banister, 2014). Urban settlements, particularly central business districts (CBDs) and heritage towns, face significant challenges due to population growth, increased urban activities, and rising vehicular movements, resulting in environmental issues like pollution and traffic congestion (Litman, 2012; Litman & Burwell, 2006). To address these social and environmental concerns, promoting non-motorized and public transportation modes is advocated as they offer efficient solutions to mobility needs (Litman, 2012; Litman & Burwell, 2006). Non-motorized and public transport modes can contribute to the sustainability and liveliness of urban environments (Litman, 2012; Litman & Burwell, 2006). In South Africa, urban areas serve as economic hubs, and their optimal functionality is vital for socioeconomic growth (Mhlanga, 2016). Developing an integrated public transport network is key to ensuring sustainable, affordable, and efficient transportation solutions for urban commuters (Mhlanga, 2016). Notable examples of rapid public transport networks contributing to economic development, job creation, and tourism include:

- MyCiTi in Cape Town
- Rea Vaya in Johannesburg

- A Re Yeng in Pretoria
- Go George in George
- Harambee in Ekurhuleni

The lack of proper institutional arrangements, such as planning, coordination, and implementation, has contributed to challenges related to non-motorized transport in South Africa (Department of Transport, 2008). The safety needs of low-income earners, who predominantly rely on walking, cycling, motorcycling, and public transport, are not adequately addressed in transport systems of both developed and developing countries (Department of Transport, 2008). To address issues of accessibility and mobility, the Department of Transport introduced the national non-motorized transport policy, recognizing non-motorized transport as a viable mode that can cater to a broader spectrum of the population (Department of Transport, 2008). Non-motorized transport includes various human-powered means such as walking, cycling, animal-drawn transport, and small-wheeled transport (Department of Transport, 2008).

Non-motorized transport offers affordability, health benefits, reduced pollution, versatility, reliability, and support for local community facilities (Department of Transport, 2008; Musakwa & Gumbo, 2017a, b; Aderibigbe & Gumbo 2022a, b). Encouraging a shift from private car usage to non-motorized transport can lead to more efficient use of existing road networks and deliver significant economic and environmental advantages (Musakwa & Gumbo, 2017a, b). In South Africa, where road transport is dominant, walking remains the primary mode of transport, especially in rural areas (Department of Transport, 2008). Although bicycles seem suitable for parts of South Africa due to flat terrain and short distances, their utilization is limited by inadequate infrastructure such as cycle paths (Musakwa & Gumbo, 2017a, b). The South African public transport industry comprises commuter rail systems, including the Gautrain high-speed rail, subsidized and unsubsidized commuter buses, bus rapid transit systems, and the minibus-taxi industry (Aropet, 2017a, b). However, these public transport systems face challenges such as poor accessibility, low ridership, congestion, and safety concerns (Saferplaces, 2017; Walters, 2014). Historically, the South African passenger transportation system primarily focused on transporting laborers, particularly those of African descent, from peripheral areas to urban centers (Khosa, 1995). Despite the provision of public transport, many commuters still experience difficulties due to long travel distances and limited access to available public transport (Thomas, 2016).

To address these issues, the South African Department of Transport has recognized the importance of non-motorized transport and introduced policies to promote alternative modes of transportation, including cycling, animal-drawn transport, and walking (Department of Transport, 2008). These policies aim to enhance accessibility, reduce travel costs, and improve the overall efficiency of transportation systems (Musakwa & Gumbo, 2017a, b). By encouraging the use of non-motorized transport, the average radius of action for individuals can be expanded, making jobs and markets more accessible (Musakwa & Gumbo, 2017a, b). In conclusion, transportation is crucial for socioeconomic development, and the provision of efficient and sustainable transport systems is essential to improve the quality of life, alleviate poverty, and enhance market integration, education, and healthcare services. Emphasizing non-motorized and public transportation modes, along with the development of supportive infrastructure, can address social and environmental concerns while ensuring accessibility and mobility for all segments of the population. However, challenges related to institutional arrangements, safety, and the optimization of public transport systems need to be addressed to achieve a more inclusive and efficient transportation network in South Africa.

Trains

According to the National Household Travel Survey (NHTS, 2018), commuters face various challenges when using train services. These include punctuality issues, as trains often do not arrive on time. Additionally, problems such as congestion, overcrowding, and security concerns act as limiting factors for train usage. The Metrorail trains, as shown in Fig. 3.9, have been particularly affected by inconsistent scheduling. Despite the provision of a timetable on their website, the NHTS (2014) findings revealed that trains frequently failed to adhere to the schedule, with 37.8% of train users stating that the timetable was unavailable. To address some of these train service issues, the Gautrain project was introduced. This project represents the largest infrastructure investment in South Africa's transport history. However, the Gautrain project has sparked public debate, with stakeholders in the transport sector raising concerns. For example, Van der Westhuizen (2007) examined the justifications for the Gautrain project, as shown in Fig. 3.10. The project was initially projected to cost 27 million Rand and was designed to cater to a small portion of the already privileged vehicle-owning population in the province. The study revealed that the project's justification was primarily driven by political

Fig. 3.9 An overview of the Metrorail train system in South Africa

Fig. 3.10 An overview of the Gautrain in South Africa. (Source: Wikipedia, 2023)

imperatives favoring regions that attracted global capital investment. As a result, the Gautrain project primarily benefited the advantaged and high-income population, perpetuating spatial and socio-economic inequalities associated with apartheid.

By developing a high-quality infrastructure that predominantly serves central locations while neglecting peripheral areas, the Gautrain project further entrenches the privileged position of the well-located, wealthier portion of the population. This exacerbates the spatial and economic segregation experienced by individuals living in peripheral regions who rely on public transport daily and require improved public transport infrastructure and services. Consequently, the Gautrain has faced significant criticism for reinforcing these inequalities.

Mini-bus taxis

In South Africa, the availability of low-capacity vehicles, specifically 16-seater mini-bus taxis, offers commuters a convenient door-to-door service with flexibility (NHTS, 2018). Compared to trains, these taxis provide a more accessible transportation option due to their adaptable routes and network (NHTS, 2018). However, there are several challenges associated with this mode of public transport, limiting its full potential for commuters. One significant issue is the operation of mini-bus taxis without licenses, and in some instances, they are driven by unlicensed drivers, which compromises safety and legality (Mtizi, 2017). This lack of regulation has led to complaints from commuters who are affected by the violence commonly associated with this transportation mode (Mtizi, 2017). Additionally, taxi drivers often exhibit reckless driving behavior, frequently violating traffic rules, and the fares charged are not consistent but vary based on peak times and weather conditions. The absence of strict laws or regulations to govern taxi drivers, particularly regarding fares, exacerbates the problem (Mtizi, 2017). Furthermore, there have been reported cases of sexual harassment and other crimes committed by both commuters and drivers, emphasizing the need for stricter law enforcement during roadblocks (Mtizi, 2017).

Buses

Buses play a vital role in public transportation systems, but several complications hinder their effectiveness. One of the primary issues is the infrequent availability of bus services during off-peak hours (Mtizi, 2017). This limitation forces commuters to seek alternative modes of transport or endure long walks to reach their destinations when buses are not operational on specific routes. Despite these challenges, buses are generally considered a safer option compared to other modes of transport. However, for the average bus commuter, these obstacles result in longer travel times and increased transport costs. Furthermore, the lack of comprehensive travel information for all modes of transport in South Africa presents an ongoing challenge. This information gap affects commuters' ability to plan their journeys efficiently and make informed decisions about their transportation options. To address these issues, there is a pressing need for greater regulation within the bus service industry. Commuters have consistently raised concerns about the lack of policies and universal guidelines that facilitate effective information sharing among the various stakeholders in the public transport sector. Figure 3.11 provides a typical example of the bus system under discussion.

Fig. 3.11 An overview of the Bus system in South Africa

In summary, the transportation modes in South Africa suffer from poor maintenance and outdated infrastructure, posing risks to both individuals and the environment. One of the drawbacks is that buses are subject to speed limits and often operate at low speeds for extended periods (Mtizi, 2017). Additionally, minibus taxi operators contribute to noise pollution through their constant calling and hooting to attract passengers. Another concerning aspect is the lack of environmental consciousness among many public transport operators, which poses a significant challenge for the future development of sustainable public transportation in South Africa.

3.11 Barriers to the Implementation and Use of Non-motorized and Public Transport in South Africa

Safety is a significant barrier to the widespread adoption of non-motorized transport (NMT) in South Africa. Policies aimed at promoting the use of cycling and other non-motorized modes of transportation often face challenges due to safety concerns, particularly for children. To encourage the substitution towards non-motorized modes, it is crucial to establish safer transport networks. Apart from concerns regarding traffic safety, the aspect of social safety also contributes to the reluctance of using non-motorized and public transportation in numerous urban areas. The perception of insufficient social safety can impede the acceptance of non-motorized transport. Local governments hold a significant responsibility in addressing this issue by integrating social safety as a fundamental consideration when planning infrastructure for non-motorized transport modes. Pedestrian safety has been overshadowed by policies that prioritize safe driving, often neglecting the needs of pedestrians (Reference 5). Consequently, pedestrians are blamed for being on roads, which are primarily viewed as the domain of motorists, disregarding the needs of cyclists and pedestrians.

In South Africa, there are several obstacles that hinder the use of non-motorized transportation (NMT), primarily influenced by geographical factors. These include challenging terrain characterized by significant water bodies, steep inclines, adverse weather conditions, land development patterns, busy roads, highways, and limited safe crossing options along railway lines. It is important to note that in South Africa, the majority of transport travel is for essential trips rather than leisure, which impacts the perception and usage of cycling as a mode of transport (Reference 7). Cycling is primarily viewed as either a recreational activity or a necessary mode of transport for those who cannot afford alternatives. Non-governmental organizations and the private sector play a significant role in promoting NMT in South Africa (Reference 8). However, the behavior of road users, including drivers, cyclists, and pedestrians, often lacks discipline, and the road environment itself may not always be conducive to ensuring the safety of public and non-motorized transport users.

3.12 Way Forward

Public transport and non-motorized transport play a crucial role in facilitating economic and societal activities. However, there is a significant disparity in the provision of efficient and affordable public and non-motorized transport for individuals with lower incomes. Addressing this challenge requires context-specific solutions in South Africa, as proposed by Mthimkhulu (2017):

1. Wide-range planning is necessary to enhance accessibility and establish an integrated transport system, particularly in rural areas. Non-motorized transport should be prioritized for people in remote locations, with a specific emphasis on utilizing bicycles. Mthimkhulu (2017) highlights that cycling is a cost-effective and environmentally friendly option that can greatly improve access for rural populations.
2. In the South African minibus taxi industry, stringent guidelines should be implemented for taxi drivers. This includes maintaining a taxi driver registry to ensure valid licenses and roadworthy vehicles. Additionally, drivers should undergo training in defensive driving and first aid (Mthimkhulu, 2017).
3. Mbara (2006) emphasizes the responsibility of local authorities in providing infrastructure and services to residents in both urban and rural areas. Proper land use planning is crucial, integrating residential and employment areas to mitigate transport costs. It is essential to develop areas alongside existing infrastructure, reducing the need for substantial capital investments in transport infrastructure.

3.13 Transport Situation in India

With the surge in traffic volume observed in India over the past two decades, there has been a heightened demand for effective traffic management systems (Zhang et al., 2011). According to Selvathi et al. (2017) and Pankaj Mukhija (2022), there has been a continuous increase in vehicle ownership in India and this increase has led to other transport problems such as accidents transportation delays, traffic congestion, and pollution emissions (Figueiredo et al., 2001). Numerous efforts have been undertaken in India, particularly in metropolitan areas, to address transportation issues; however, none have entirely resolved the persistent challenges faced by most cities. Singh (2005) argued that expanding road capacity through widening or constructing signal-free or new roads is not cost-effective in India. Consequently, the transportation system grapples with various challenges such as inadequate lighting, adverse weather conditions, and road quality concerns.

The automotive sector plays a significant role in India's Gross Domestic Product (GDP), with both the Government of India and State Governments, alongside various automobile-related entities, emphasizing the adoption of smart technology and Intelligent Transportation Systems (ITS). These systems encompass a range of innovations such as smart cards, incident response mechanisms, advanced driving assistance, and real-time parking management, as highlighted by Pathak et al. (2021). Transportation in India holds immense economic importance, serving as a vital component for businesses and industries. With India's extensive 4.7 million km of road network, the transport sector handles approximately 65% of freight and 80% of passenger traffic, further underlining its pivotal role in the nation's economy. Despite this, the road transport infrastructure is poor and inadequate, thus, making safety a matter of concern (Grant Thornton, 2016). In addition to the above, the persistent rise in population and per capita income, which is a major contributor to traffic congestion times, which leads to loss of valuable manhours to the nation, also affects the environment through the reliance on vehicles that utilizes fossil fuel and hence leads to increased carbon emission. As a result of this, it is very much necessary to address the transport situation in the country and suggest strategies such as the adoption of technology in resolving the problem.

3.14 Transport Problems in India

Inadequate Public Transport Infrastructure: Indian cities grapple with a significant challenge posed by an inefficient or redundant public transport system. The lack of substantial measures to enhance it has resulted in a substantial increase in private vehicle ownership, consequently escalating the demand for additional transport infrastructure to address traffic congestion. Despite the government's projection in 2007 of a 3% growth in vehicular numbers, the actual increase stood at 7%, as highlighted by the World Bank (2017).

3.14.1 Inadequate Infrastructure

As per a report by Truck Guru (2022), roads constitute 60% of the overall transportation network. In India, road transit emerges as the predominant mode of transportation. Nonetheless, the challenges confronting road transport in India are escalating significantly over time, impacting transportation firms due to inadequate road infrastructure, numerous checkpoints, and congestion. Whether conveying large cargo or delicate goods, a well-maintained road network is indispensable to ensure the safe delivery of commodities to their destination. The primary issue concerning India's transport infrastructure is the prevalence of outdated and deteriorating facilities across all modes of transport. In the railway sector, approximately 25% of route lengths and 75 to 80% of machinery in railway workshops are in poor condition and require replacement. Similarly, more than 80% of public buses operated by State Road Transport Corporations are outdated and obsolete, as highlighted by Mehta (2021).

3.14.2 Poor Traffic Management

As per traffic reports from leading nations worldwide, India, being the most populous country, faces a myriad of traffic challenges and potential solutions. The burgeoning urban population has resulted in extended periods spent in traffic congestion for both freight vehicles and travelers alike.

3.14.3 Inadequate Parking Facilities

Parking poses a significant challenge in India's transportation system. Illustrated in Figure ..., there is a notable shortage of parking facilities in the central areas of large cities to accommodate motorists adequately. Consequently, parking management becomes necessary, often implemented by restricting commuters from parking all day or by imposing high parking fees. However, in some towns, the fines for parking violations are so nominal that repeat offenders find it more economical to pay fines multiple times rather than adhere to parking regulations, resulting in minimal deterrence for offenders (Fig. 3.12).

Fig. 3.12 View of a traffic and parking condition in an India City. (Source: Truck Guru, 2022)

3.14.4 *Poor Planning of Transport System*

The uneven development of the transport system within the country results from the lack of integration among various modes of transport, placing significant strain on rail and road networks in specific cities and regions. To achieve balanced regional development, the establishment of alternative routes is imperative; for instance, the introduction of the Metro in Delhi has alleviated pressure on road transport. Moreover, there is a crucial need for the integration of all transportation modes to mitigate reliance on a singular mode. The absence of such integration fuels the proliferation of personal transport, exacerbating pressure on roads and contributing to transportation challenges such as traffic congestion, delays, and accidents.

3.15 Motorised and Public Transport Use in India

Walking, cycling, and riding a cycle rickshaw exemplify non-motorized modes of transportation. Cycle rickshaws serve as vital intermediate para-transit in numerous Indian cities (IPT). These modes operate without the use of fossil fuels and produce minimal pollution, making them genuinely low-carbon alternatives. However, both public transportation and non-motorized transport (NMT) have experienced declines in urban areas, leading to increasing negative externalities. The National Action Plan on Climate Change (NAPCC) of India acknowledges that reducing transportation-related greenhouse gas (GHG) emissions can be achieved by encouraging low-income households to rely on various transportation modes for commuting to work, school, and accessing essential services. Despite the health benefits associated with non-motorized transport (NMT), its utilization has declined due to growing affluence and inadequate infrastructure (Singh & Dev, 2023). Presently, non-motorized transport (NMT) and public transport (PT) hold a significant modal share in Indian cities, although historical data indicates a decline in their usage. The substandard quality of existing NMT and PT infrastructure in Indian cities exacerbates the risk of road traffic accidents for these users. Consequently, there exists a likelihood that current NMT and PT users will transition to personal motorized vehicles (PMV) once they become financially viable options. However, retaining and potentially increasing the share of NMT and PT users is feasible through the establishment of safe and convenient facilities tailored to their needs, as suggested by Tiwari et al. (2016).

Historical trends indicate a consistent decline in the utilization of Non-Motorized Transport (NMT) and Public Transport (PT) in Indian cities. The current infrastructure for NMT and PT in these cities is in poor condition, posing discomfort and high risks of traffic accidents for users. Moreover, despite the decline in the percentage of individuals utilizing non-motorized transport such as bicycles and human-powered vehicles, the absolute numbers remain substantial, as noted by Tiwari & Jain (2008). It is estimated that even though cycle trips in major cities may represent only 6–8% of total trips, in absolute terms, they could exceed a million commuter trips, with a significant portion of users belonging to low-income groups who lack the means to afford private automobiles for mobility. While many Indian cities are investing in initiatives aimed at enhancing vehicle mobility, such as road widening schemes and the construction of flyovers and foot-over bridges, minimal efforts are directed towards improving basic transport infrastructure for pedestrians, cyclists, and bus users. Given the prevailing conditions, users of low-carbon transportation modes are likely to transition to carbon-intensive modes once they can afford to do so, as indicated by Tiwari & Jain (2013) as cited in Tiwari et al. (2016).

3.16 Challenges to the Adoption and Effective Utilization of NMT and PT in India

In the 1980s, the utilization of non-motorized transportation (NMT) in Indian cities exhibited a correlation with income levels, as noted by Replogle (1991). However, as income levels rose, there was a decline in the use of both NMT and public transportation (PT). This decline can be attributed to the inadequate investment by city and state governments in NMT and PT infrastructure, resulting in diminished service quality and heightened risks for pedestrians and cyclists. Despite this trend, NMT continues to dominate as the primary mode of transport in Indian cities, even in megacities with populations exceeding 8 million people, where walking and bicycling collectively account for a median share of 40–50%. A study conducted by Singh and Dev (2023) revealed that bicycle ownership rates are high across all Indian cities. According to the 2001 Census, between 35 and 65% of households in most medium and large cities possess one or more bicycles, with ownership rates varying between 33 and 48% in smaller cities (except for Mysore, where only 27% of households own bicycles). Notably, 54.43% of households in Ahmedabad and 63.4% in Chandigarh own one or more bicycles, while 37.6% of Delhi households owned bicycles in 2001 (Tiwari & Jain, 2013). Furthermore, a study by The Energy and Resources Institute (TERI) titled 'Benefits of cycling in India' reported that by 2015–16, approximately 23% of working individuals in urban areas relied on walking, while 13% depended on bicycles for their daily commute, indicating that 36% of the urban working population utilized non-motorized transportation systems for commuting purposes.

The quality of infrastructure and its availability were some of the hindering factors preventing the use of non-motorised and public transport in some cities in India. Several factors such as speed, travel time, accessibility, were used to evaluate NMT infrastructure in Delhi, Vadodara, Jaipur, and Patna in order to uncover the challenges facing commuters during active mobility. It has been observed that pedestrian infrastructure, including pathways, is either absent or poorly maintained, resulting in direct conflicts between pedestrians and motorized traffic. Bicyclists encounter difficulties in utilizing designated infrastructure due to inadequate design and a lack of continuous networks in specific areas (Singh & Dev, 2023). Moreover, the available infrastructure is not seamlessly integrated with other public transportation systems, and in certain locations, it is entirely absent, disregarding the needs of low-income populations heavily reliant on Non-Motorized Transport (NMT) (Tiwari et al., 2016). In cities like Chennai and Bengaluru, the evaluation of NMT infrastructure considers factors such as design, speed, travel time, accessibility, safety, availability of cycle lanes, riding surface, and street connectivity. Pedestrian infrastructure suffers from similar shortcomings, including non-existent, damaged, or obstructed pathways, significant height disparities between sidewalks and driveways/alleyways, flooding during rainy seasons, hazards at roadway crossings, and inadequate spacing between crosswalks. Furthermore, footpaths lack continuity due to disruptions caused by utility repairs and maintenance, and few exclusive-use lanes for bicyclists are available, leading to their displacement from major highways by motor vehicles on adjacent roadways (Bank, 2005; Gota et al., 2013).

Furthermore, the assessment of Non-Motorized Transport (NMT) infrastructure in Delhi includes factors such as the presence of cycling lanes, quality of riding surfaces, and street connectivity. Presently, approximately 40% of the city's roads lack pedestrian pathways, and where they do exist, they are often inaccessible, particularly for individuals with disabilities. Zebra crossings are frequently ignored, and controlled crossings are spaced at considerable distances. There is a growing emphasis on facilitating unhindered movement for motorized vehicles through the construction of foot over bridges and subways (Mamun et al., 2009; Singh, 2018). It has also been observed that pedestrians and cyclists have limited infrastructure options, as current transportation development initiatives in India prioritize rapid modes of transport that claim Right of Way (ROW) by displacing existing facilities.

3.17 Development of Smart/Intelligent Transportation System (ITS) in India

The tremendous growth in information and communication technology infrastructure has seen an improvement in the transport sector globally. The study of Oluwayemi-Oniya Aderibigbe and Trynos Gumbo (2024) asserted that the role of information and communication technology has complemented majority of household trips, hence, reducing the emission of carbon emission into the atmosphere. It has also built a strong base for intelligent transport systems by connecting the objects on the roads in a reliable and secure way (Bonomi, 2013).

Pankaj Mukhija (2022) highlighted the developments of ITS in India into four stages or four generations. These are:

A. First Generation ITS: The First Generation of Intelligent Transportation Systems (ITS) in India spanned from 1950 to 1980, focusing primarily on developing essential highways and roads (Kesharwani, 1976). During this period, technological advancements in India were limited, and efforts were concentrated on establishing national and state routes. Rural road connectivity was in its nascent stage, and automated systems were primarily confined to metropolitan areas.

This era of ITS can be characterized as infrastructure-based. The introduction of automated traffic light signals marked the initial steps towards an intelligent transportation system. Many ITS concepts were still in the research phase during this time. However, in the late 1980s, advancements in communication technologies, microprocessors, and microcontrollers led to the incorporation of additional features like traffic diversion and route guidance into ITS.

B. Second Generation ITS: The Second Generation of Intelligent Transportation Systems (ITS) in India spans from 1980 to 1995. During this period, the Government of India and private industries began collaborating on ITS concepts. This generation saw the introduction of two-way communication features in national highway projects, as well as enhancements in toll collection systems and parking management services. Advancements in computer technology facilitated the integration of additional applications into ITS. However, these developments were primarily concentrated on national highways and in a few urban areas, with limited progress in ITS implementation in rural India.

C. Third Generation ITS: The third generation of Intelligent Transportation Systems (ITS) in India spans from 1995 to 2010. This phase saw the introduction of automatic and interactive vehicle operations. The primary focus during this period was on aligning the interests of various stakeholders, developing cooperative systems within ITS, standardizing activities to prevent conflicts, and promoting commercial deployment. Key technologies that emerged in this phase included Mobile Ad Hoc Networks (MANET) and Vehicular Ad Hoc Networks (VANET). Additionally, more sophisticated features such as the Emergency Management system were integrated into ITS, and numerous ITS projects were initiated.

D. Fourth Generation ITS: The fourth generation of Intelligent Transportation Systems (ITS) in India spans from 2010 to the present. This era has seen significant advancements due to technological innovation, the integration of diverse technologies, and extensive work in the field of the Internet of Things (IoT). These developments have led to exponential growth in India's transport system, with numerous features being incorporated into ITS. Real-time applications increased in this period, and specific solutions for individuals and localities were considered in this phase. More traffic management models were developed and more features like Advanced Driver Assistance, an intelligent system for pedestrian crossing, a real-time arrival information system of vehicles, E-Challan, variable message signs, and smart traffic lights were included (Ravindra, 2020).

3.18 Existing Smart Transport System in India

3.18.1 Modern Age Vehicles

In India today, many vehicle manufacturers are incorporating advanced technological features that enhance time management and rider safety. These modern vehicles are designed for superior performance and efficient fuel consumption. Additionally, most come equipped with Internet connectivity and Internet of Things (IoT) capabilities, facilitating improved transportation and adherence to Bharat Stage and Euro standards. Safety enhancements include crash testing certifications, the use of high-quality metal for the car body, Anti-lock Braking Systems (ABS), and airbags.

3.18.2 *Use of Electric Cars*

While petrol and CNG vehicles dominate the Indian market, manufacturers like Mahindra and Toyota have introduced electric cars. However, the widespread adoption of electric vehicles (EVs) faces challenges, primarily due to inadequate charging infrastructure, which hinders effective EV use in many cities in the Global South. To address this, the Indian government has launched the National Electric Mobility Mission Plan 2020. This plan incentivizes the adoption of green vehicles and promotes domestic manufacturing capabilities in the automobile sector by linking it to the Make in India scheme (Pathak et al., 2021).

3.18.3 *Internet of Things (IoT)*

According to Pathak (2021), although the adoption rate of IoT among Indian citizens, particularly in transportation, is currently low, there is expected to be a rise in the popularity of OnBoard Diagnostic (OBD) port devices in the automobile sector. These devices, installed in cars and connected to cellular networks, provide users with location data services and automotive telematics. They enable cars to communicate with each other, aiding in better traffic management through the creation of real-time traffic data. Reliance Jio is working on bringing such devices to market, which will be constantly connected to its LTE network. Additionally, these devices will offer users a mobile application providing vital information such as oil status, water levels, fuel information, and battery change alerts. Remote commands can also be sent to the car, such as turning ON the AC from a mobile phone, ensuring the cabin is cool by the time the user reaches the vehicle.

3.19 Future Direction Towards Intelligent Transport System in India (ITS)

India's Intelligent Transportation System (ITS) leverages Information and Communication Technology (ICT) to manage transportation efficiently. The system comprises both backend and hardware components, offering solutions such as driver dashboards, electronic ticketing machines, passenger information systems, and vehicle tracking systems. Key areas where ITS is emerging in India include:

3.19.1 Boom in Communications and Analytics Infrastructure

India is capitalizing on the widespread use of smartphones, now largely powered by LTE network infrastructure that covers 99% of the population. Applications that collect location-based data can analyze real-time information, which, when combined with services like weather updates and shopping/work hours, provides users with efficient transportation information. This development is bolstered by the growth of data virtualization and cloud-based analytical tools, ensuring ubiquitous access to transportation data and insights.

3.19.2 Radio Frequency Identification (RFID)

This is especially significant in freight transportation. While RFID technology is not new to India, its full potential has not been realized in the freight sector, even though it is widely used by global counterparts. This is despite the logistics sector experiencing rapid growth in India. According to Pathak (2021), India ranks 46th among 155 countries in the World Bank's International Logistics Performance Index. Recently, Indian Railways has decided to implement RFID technology in its freight coaches. These devices will enhance freight maintenance management, locomotive management, and coach maintenance by ensuring more effective, transparent operations and real-time tracking.

3.19.3 Electronic Toll Payments

Electronic toll collection systems use RFID technology to read data from an RFID tag on a vehicle from a distance, automatically deducting the toll and opening the barrier at each entry point. This technology saves citizens time and fuel at toll gates, and it improves traffic management by reducing travel time and congestion. The National Highways Authority of India (NHAI) has introduced the FASTag service nationwide to facilitate this process. FASTag is linked to a prepaid wallet from which the toll amount is deducted, and it uses RFID technology, with the tag placed on the vehicle's windscreen. FASTag is currently operational at most toll plazas on state and national highways. The adoption of such devices also benefits industries that manufacture semiconductor chips, RFID devices, and related IT systems.

3.19.4 Unified Smart Cards for Public Transport

Smart card usage has already begun in urban areas like metro train services. The Delhi Government and various authorities in Mumbai are considering the implementation of integrated smart cards. These RFID-based cards enable citizens to pay for multiple types of public transportation, including metro, local trains, and buses, using a single card. This facilitates seamless multimodal transportation systems in major cities. The unified smart cards for public transport promote the widespread adoption of smart technologies, enhancing public transportation in India and creating significant opportunities for the semiconductor industry.

3.19.5 Passenger Information System

Expected time of arrival (ETA) is shown on electronic signboards at bus stops and train platforms. These systems also deliver information to personal mobile devices. For instance, the DIMTS "Poocho" app for Delhi buses provides the ETA for buses on specific routes, reducing uncertainty and congestion at waiting areas. Additionally, these display systems can generate revenue for organizations by displaying advertisements.

3.19.6 Use of Live Traffic Data

With the extensive communication capabilities offered by cellular technology and the widespread use of smartphones, collecting traffic data has become more accessible. This data can be utilized to anticipate delays in specific routes and offer alternative routes in real-time, promoting efficient traffic flow and consequently reducing carbon emissions. Hyderabad is poised to undergo a transformation in its traffic management system in the next decade, with the approval of the Citywide Intelligent Transportation System (ITS) initiative by the Government, allocated a budget of 1180 crores. In conclusion, transport problems in India is no different from what is experienced in other countries in the global south, hence, the role of technology such as the application of intelligent transport as well as other sustainable means in combating these problems can be sidelined.

In conclusion, policy interventions and improvements in transport systems are necessary in developing countries such as Nigeria and South Africa to enhance their competitiveness compared to developed nations. This entails adequate investment in transport infrastructure and the integration of multiple transport modes to reduce reliance on a single mode, promoting sustainable planning and development.

References

Ademiluyi, I., & Gbadamosi, K. (2004). Geography of movement and mobility crisis in Nigeria. *International Journal of Environment and Development, 7*(1), 22–29.

Aderibigbe, O. (2023). Rural-urban connectivity: A way of Achieving sustainable rural transport development. In *Handbook of research on managing the urban-rural divide through an inclusive framework* (pp. 285–301). IGI Global. ISSN: 2475-675X; eISSN: 2475-6768. https://doi.org/10.4018/978-1-6684-6258-4.ch016

Aderibigbe, O. O., & Gumbo, T. (2022a). Variations in mode choice of residents prior and during COVID-19: An empirical evidence from Johannesburg, South Africa. *Sustainability, 14*, 16959. https://doi.org/10.3390/su142416959

Aderibigbe, O. O., & Gumbo, T. (2022b). Influence of socioeconomic attributes on travel behaviour in the rural areas of Nigeria: Towards a sustainable rural planning and development. *Urban, Planning and Transport Research, 10*(1), 181–199. https://doi.org/10.1080/21650020.2022.2072946

Ali, A.K.A.K. (2013). Explaining Smart Growth applications: lessons learned from the US capital region. [online]. *Urban Studies, 51*(1), 116–135. SAGE Publishers. Available from:. [Accessed on: 26 February 2018].

Anyanwu, J. C. A., Oyefusi, H., Oaikhenan, F., & Dimowo, A. (1997). *The structure of the Nigerian economy (1960–1997)*. Joanee Educational Publishers Ltd.

Aropet, A. (2017a). Public transport modes and challenges in South Africa: A review. *Journal of Transportation Technologies, 7*(2), 135–149.

Aropet, R. (2017b). Southern African solutions to public transport challenges. In *36th Southern African transport conference* (SATC 2017). ISBN:978-1-920017-73-6.

Bank, T. H. E. W. (2005). Transport economics, policy and poverty thematic group. *Reading, 1*, 1–7.

Belwal, R., (2016). Multimodal integrated smart sustainable transport system – UITP MENA. [online]. *Center for Transport Excellence Journal*. Accessed 24 Aug 2017.

Bonomi, F. (2013). The smart and connected vehicle and the Internet of Things. In Invited Talk, Workshop on Synchronization in Telecommunication Systems.

Brussel, M. (2014). Integration of public transport and NMT-UNEP. Principles and application in an East African context. [online]. University of Twente PDF Presentation. Available from: http://staging.unep.org/transport/sharetheroad/PDF/courseware_nmt/Lecture8_Integration_Publictransport_NMT_Brussel.pdf. Accessed 12 June 2017.

Buhari, M. (2000). The role of infrastructural development and rehabilitation in sustainable economic growth in Nigeria. In A paper presented at the all peoples party economic summit, held at the Ladi Kwali conference Center, Sheraton Hotel and Towers, Abuja, 9–10 November, 2000, http://www.buhari2003.org//speeches.htm.

Connor, G. (1993). Developing mass transit infrastructure and traffic facilities. In S. Ikya (Ed.), *Urban passenger transportation in Nigeria*. Heinemann Educational Books (Nig.) Plc.

Department of Transport. (2008). *Draft national non-motorized transport policy: Transport, the heartbeat of South Africa' economic growth and social development*. http://www.transport.gov.za/. Accessed 21 Jan 2022.

Dosunmu, V. A., & Mogbojuri, O. (2019). Perspective of integrated transport system for sustainable development in an emerging economy. *International Journal of Communications and Network, 2*, 8. ISSN:2637-6571.

Fawcett, P. (2000). *Managing passenger logistics: The comprehensive guide to people and transport*. Kogan Limited.

Federal Republic of Nigeria. (1995). *National urban transport policy for Nigeria*. Draft Policy Document, Federal Urban Mass Transit Agency, Abuja.

Federal Road Safety Cooperation (FRSC). (2011, March 31). Successes and challenges of a lead agency and the multisectoral nature of road safety: A presentation by Osita Chidoka Corps Marshal and Chief Executive, Federal Road Safety Corps.

Figueiredo, L., Jesus, I., Tenreiro Machado, J. A., Rui Ferreira, J., & Martins De Carvalho, J. L. (2001). Towards the development of intelligent transportation systems. In *IEEE proceedings of conference on intelligent transportation systems* (pp. 1206–1211). ITSC. https://doi.org/10.1109/itsc.2001.948835

Filani, M. O. (2002). Mobility crisis and the federal government's mass transit 13 programme. In S. O. Onakomaiya & O. O. Oyesiku (Eds.), *Environment, physical planning and development in Nigeria* (pp. 37–51). Olabisi Onabanjo University.

Filani, M. O. (2003). *Advancing the cause of private participation in the road transport sub-sector in Nigeria*. A paper delivered at the 10th anniversary celebration of the Associated Bus Company (ABC) Ltd., on May 14, 2003 at Ikeja.

Glen, W., & Reno, A. (2009). *Economic Impact of Public Transportation Investment*. American Public Transportation Association. TCRP Project J-11, Task 7, Transit Cooperative Research Program.

Gota, S., Anthapur, S. K., & Bosu, P. (2013, December 12–21). *Promoting non-motorized transport in Asian cities: Policymakers' toolbox*.

Grant Thornton. (2016). *Smart transportation: Transforming Indian cities*. Retrieved from http://www.grantthornton.in/globalassets/1.-member-firms/india/assets/pdfs/smart-transportation-report.pdf

Heggie I. (1995). *Management and financing of roads*. No 275, world bank report.

Jain, D., & Geetam, T. (2020). Gender and income based variability in travel choices in Vishakhapatnam, India. *Transportation Research Procedia, 48*, 2870–2890. https://doi.org/10.1016/j.trpro.2020.08.232

Johnson, C., & Johnson, D. (2019). Promoting active travel in South Africa: The role of non-motorized transport infrastructure. *Journal of Transport Geography, 74*, 336–345.

Joshi, M. S., & Maclean, M. (1995). Parental attitudes to children's journeys to school. *World Transportation Policy and Practice 1*(4), 29–36. https://doi.org/10.1108/13527619510102016

Kesharwani, T. R. (1976). Development of transportation in India: A broad perspective. *Economic and Political Weekly, 11*(26), 960–964.

Khosa, M. N. (1995). The urban transport system in South Africa: An overview. *Journal of Transport Geography, 3*(1), 45–53.

Litman, T. (2012). *Evaluating non-motorised transportation benefits and costs*. Victoria Transport Policy Institute.

Litman, T., & Burwell, D. (2006). Issues in sustainable transportation. *International Journal of Global Environmental Issues, 6*(4), 331–347.

Mabogunje, A. L. (2002). Re-constructing the Nigeria city: The New policy on Urban Development and Housing. A keynote address in Amole, D. The city in Nigeria: Perspectives, issues, challenges, strategies. *Proceedings of a National conference organized by the Faculty of Environment Design and Management*. Obafemi Awolowo University, Ile-Ife, Nigeria, pp.1–9.

Mamun, M., Este, D., & Non, J. M. (2009). Non-Motorized Public Transport Development Present Scenario & Future Approach in Developing Cities. Proceedings for the 3rd Smart Systems Postgraduate Student Conference, October. https://eprints.qut.edu.au/

Marais, L. (2020). Advancing transportation planning in South Africa: Strategies for developing world-class cities. *Urban Planning, 5*(4), 1–12.

Marshall, S., & Banister, D. (2014). *Land use and transport: European research towards integrated policies*. Elsevier.

Mbara, T. C. (2006). *Coping with demand for urban passenger's transport in Zimbabwe: Challenges and options*. University of Zimbabwe.

Mehta, P. (2021). *Main problems of transport development in India*. https://www.economicsdiscussion.net/articles/main-problems-of-transport-development-in-india/2183. Assessed November.

Mhlanga, O. (2016). Urban transport policy implementation in South Africa: Challenges and prospects. *Journal of Transport and Supply Chain Management, 10*(1), a222.

Mthimkhulu, N. (2017). Southern African solutions to public transport challenges. In *36th Southern African Transport Conference (SATC 2017)*. ISBN: 978-1-920017-73-6.

Mtizi, C. (2017). Southern African solutions to public transport challenges. In *36th Southern African Transport Conference (SATC 2017)*. ISBN: 978-1-920017-73-6.

Musakwa, W., & Gumbo, T. (2017a). Factors affecting travel behavior and non-motorized transport use: A case of residents in Cape Town, South Africa. *Journal of Transport Geography, 60*, 142–150.

Musakwa, W., & Gumbo, T. (2017b). Impact of urban policy on public transportation in Gauteng, South Africa: Smart or Dumb City systems is the question. In R. Álvarez

Fernández, S. Zubelzu, & R. Martínez (Eds.), *Carbon footprint and the industrial life cycle: From urban planning to recycling* (pp. 339–356). Springer. https://link.springer.com/chapter/10.1007/978-3-319-54984-2_16

National Household Travel Survey 2014. (2018). Statistics South Africa.

National Transport Policy for Nigeria. (2003). Draft document. Transport Sector Reform Committee Bureau of Public Enterprises, Abuja.

NBS. (2008). National Bureau of Statistics, Annual Abstract of Statistics.

NBS. (2012). National Bureau of Statistics, Annual Abstract of Statistics.

Odufuwa, B. O. (2003). Transport poverty and the role of urban governance in Nigeria. *African Journal of Social and Policy Studies, 3*(1 and 2), 144–149.

Olomola, A. S. (2003, December). Understanding poverty in Nigeria: Highlights from NISER review of Nigerian Development. NISEREEL. *The Magazine of the Nigerian Institute of Social and Economic Research*, Nos. 4 & 5, Ibadan.

Oluwayemi-Oniya Aderibigbe & Trynos Gumbo. (2024). Impact of information and communication technology on mobility of urban and rural households: A comparative study from Nigeria. *Urban, Planning and Transport Research, 12*(1), 2348551. https://doi.org/10.1080/21650020.2024.2348551

Osinubi, T. S. (2003). Urban Poverty in Nigeria: A case study of Agege Area of Lagos State Nigeria. In *Paper presented at the World Institute for Development Economics Research (WDER) conference*, Finland.

Oye, A., Kadom, S., & John, K. (2016). Evaluating the role and impact of railway transport in the Nigerian economy, options and choices: Case of Nigerian Railway Corporation. AshEse. *Journal of Economics, 2(4)*, 103–113. ISSN 2396-8966.

Oyesiku, O. (2002, August 27). From Womb to tomb, 24th Inaugural Lecture, Olabisi Onabanjo University, Ago-Iwoye, Tuesday.

Oyesiku, O. (2004). Policy directions in urban transportation. In I. Chikolo, A. A. Ogunsanya, & A. G. Sumaila (Eds.), *Perspectives on urban transportation in Nigeria* (pp. 171–202). NITT.

Pankaj Mukhija, P. (2022). Development in the intelligent transportation system in India. *International Journal of Recent Technology and Engineering (IJRTE)*. https://doi.org/10.35940/ijrte.D7324.1111422

Pathak, P., Singhal, A., & Bhutani, S. (2021). Intelligent transportation system in India. *Telecom Business Review: SITM Journal*. file:///Users/macbook/Downloads/Intelligent_Transportation_System_in_India20190923-80370-usf15h-with-cover-page-v2.pdf. Assessed 22 Oct 2023.

Ravindra, A. I. (2020). Smart Crosswalk System. vol. 6, no. 8, pp. 23–28, 2020.

Replogle, M. (1991). *Non-motorized vehicles in Asian cities* (p. 162). World Bank. http://documents.worldbank.org/curated/en/191771468770663924/pdf/multi-page.pdf

Road Vision. (2000). *Transport in Nigeria in 2020* (p. 4). Steering Committee, Information Brochure.

Rodrigue, J., Comtois, C., & Slack, B. (2013). *The geography of transport systems* (3rd ed.).

Saferplaces. (2017). *Integrated transport planning guideline for urban areas in South Africa*. Saferplaces.

Saferspaces. (2017). *The state of public transport in South Africa*. https://www.saferspaces.org.za/understand/entry/the-state-of-public-transport-in-south-africa. Accessed 20 Dec 2021.

Salau, A. T. (1992). Urbanization and Spatial Strategies in West African. In Potter, R. and Salau, A. (Eds). *Cities and Development in the Third World*. Mansell Publishing Ltd, London. pp. 157–171.

Selvathi, D., Pavithra, P., & Preethi, T. (2017). Intelligent transportation system for accident prevention and detection. In *Proceedings of the 2017 international conference on intelligent computing and control systems (ICICCS 2017)* (Vol. 2018, pp. 442–446). https://doi.org/10.1109/ICCONS.2017.8250761

Singh, S. (2005). Review of urban transportation in india. *Journal of Public Transportation, 8*(1), 79–97. https://doi.org/10.5038/2375-0901.8.1.5

Singh, Y. P. (2018). Impact and Strategies for Slow Moving Vehicles : Case Study of Azadpur. February, 542–547. https://doi.org/10.13140/RG.2.2.13378.68802

Singh, S., & Dev, M. (2023). Challenges of successful non-motorized transport infrastructure in Indian cities. *Transportation Research Procedia, 72*(2023), 1982–1989. https://doi.org/10.1016/j.trpro.2023.11.679

Smith, A., Jones, B., & Johnson, C. (2017). Enhancing urban mobility: Lessons from the Bus Rapid Transit (BRT) and high-speed rail initiatives in South Africa. *Transportation Research Procedia, 25*, 453–460.

South African Institution of Civil Engineering (SAICE) Report (2017). SAICE 2017 Infrastructure Report Card for South Africa. Assessed on June 2023 from https://saice.org.za/wp-content/uploads/2017/09/SAICE-IRC-2017.pdf

Statistics South Africa. (2014). *National household travel survey 2013: Statistical release*. Statistics South Africa P0320.

Thomas, R. (2016). The crisis of public transport in South Africa. *Research in Transportation Economics, 59*, 21–29.

Tiwari, G., & Jain, H. (2008). 'Bicycles in urban India' in position papers on bicycling in Asia. *Interface for Cycling Expertise*, 3–20.

Tiwari, G., & Jain, D. (2013). *NMT infrastructure in India: Investment, policy and design*. Promoting Low Carbon Transport in India.

Tiwari, G., Jain, D., & Ramachandra Rao, K. (2016). Impact of public transport and non-motorized transport infrastructure on travel mode shares, energy, emissions and safety: Case of Indian cities. *Transportation Research Part D: Transport and Environment, 44*, 277–291. https://doi.org/10.1016/j.trd.2015.11.004

Truck Guru. (2022). *General problems in transportation system of India and its solutions*. https://truckguru.co.in/blog/problems-in-transportation/. Assessed Jan 2024.

Tumwebaze, K. K. (2017). Rural transport in Africa: characteristics, challenges, and opportunities. In K. K. Tumwebaze (Ed.), *Rural transport in Africa*.

United Nations Report: NMT Policies in Nigeria. file:///Users/macbook/Desktop/book%20oniya%20and%20gumbo/NMTStrategy_Nigeria_200402%20use%20well.pdf. Accessed 2 Feb 2023.

Van der Westhuizen, J. (2007). *Glits, glamour and the gautrain: mega-projects as political symbols*.

Walters, J. (2014). Public transport policy implementation in South Africa: Quo Vadis? *Journal of Transport and Supply Chain Management, 8*(1). https://doi.org/10.4102/jtscm.v8i1.134

Wikipedia. (2023). *Gautrain*. Wikipedia. https://en.wikipedia.org/wiki/Gautrain. Accessed 18 Feb 2023.

World Bank. (2017). Statistical data on transportation system by World Bank. (n.d). Retrieved from http://databank.worldbank.org/data/reports.aspx?source=2&Topic=16

Zhang, J., Wang, F. Y., Wang, K., Lin, W. H., Xu, X., & Chen, C. (2011). Data-driven intelligent transportation systems: A survey. *IEEE Transactions on Intelligent Transportation Systems, 12*(4), 1624–1639.

Part II
Sustainable Development, Smart Cities and Emerging Technologies

Chapter 4
Sustainable Transport Planning

Oluwayemi-Oniya Aderibigbe and Trynos Gumbo

4.1 Introduction

4.1.1 General Overview of Sustainability

The concept of sustainable development has been a topic of discussion, with various perspectives and definitions. From an economic standpoint, Barbier (1989) provided his definition. However, the widely accepted definition, as proposed by the World Commission on Environment and Development (WCED, 1987), is that sustainable development entails meeting the needs of the present generation without compromising the ability of future generations to meet their own needs (Drexhage and Murphy 2010). Similarly, Munasinghe (2004, as cited in Abubakar, 2013a, b) defines sustainable development as a process that enhances opportunities for individuals and communities to achieve their aspirations and full potential over a sustained period while ensuring the resilience of economic, social, and environmental systems.

Sustainable development rests upon three pillars: economic development, social equity, and environmental protection. Ayeni (2010a, b) argues that these three pillars must be considered holistically, planned collectively, and implemented in conjunction to attain the desired outcomes. The Fifth EC Environmental Action Programme (February 1, 1993) identifies key characteristics of sustainable development, which include the maintenance of overall quality of life, continued access to natural resources, avoidance of lasting environmental damage, and development that meets present needs without compromising future generations' ability to meet their own needs (M3 Future, 2005). The concept of sustainable development encompasses three dimensions: environment, economy, and society, as discussed by Theis (2012) and Black (2010). Low (2003a, b) provides a comprehensive explanation of these dimensions.

O.-O. Aderibigbe et al., *Emerging Technologies for Smart Cities*,
https://doi.org/10.1007/978-3-031-66943-9_4

- **Environment**: The environmental dimension focuses on the impact of human activities and developments on local and global environments.
- **Economy**: The economic dimension pertains to the growth and progress of a community towards economic goals such as wealth, employment, productivity, and welfare.
- **Social:** The social dimension of sustainability involves issues of equity and inclusion (Low, 2003a, b).

In line with these dimensions, Age, (2005), as cited in Abubakar, (2013a, b) identified several objectives of sustainable development. These include increasing capital incomes, employment opportunities, promoting human welfare, satisfying basic needs, and protecting the environment. Considering the crucial role of transportation in achieving sustainable development, it becomes evident that transportation, as an element of the environment, plays a significant role. As cities have expanded throughout the twentieth century, the development of transportation networks has both facilitated urban growth and presented challenges to achieving transport sustainability. The rise in urbanization, living standards, and rapid economic development has led to the growth and advancement of urban and intercity transportation systems. Consequently, many urban centers face various transportation issues such as traffic congestion, accidents, inadequate infrastructure, and insufficient maintenance policies by governments and stakeholders.

To ensure environmental sustainability, relevant stakeholders must adopt measures aimed at mitigating the risks associated with transportation management in major cities worldwide. One pressing challenge is the prevalence of automobile-centric development. Urban centers worldwide suffer from congested roads as reliance on automobiles as the primary mode of transportation increases (Moavenzadeh et al., 2002a, b). Moreover, the construction of large automobile-oriented freeways has led to the segregation of communities and contributed to social issues, while car emissions from these freeways have negative local and global environmental impacts (Banister, 2005). These consequences stem from the rapid development of transportation systems in the twentieth century, where urban planning was primarily designed to accommodate automobiles as the dominant mode of transportation (Newman & Kenworthy, 1999). The resulting congestion and automobile dependency lead to increased emissions and pollution, adversely affecting human health and hindering economic growth. These challenges highlight the overarching problem of unsustainable transportation systems (Banister, 2005). With the ongoing urbanization in developing countries and the widespread dominance of automobiles in developed nations, it becomes imperative to examine transportation policies that foster a sustainable and high-quality urban life for residents. Gumbo and Moyo (2020) underline the significance of sustainable transport as a crucial component of urban development, emphasizing the need to promote it in order to accomplish the objectives set forth by Sustainable Urban Mobility Plans.

The Agenda 2030 for Sustainable Development lays out a comprehensive framework for fostering a sustainable, prosperous, and peaceful global future. Central to this agenda are the 17 Sustainable Development Goals (SDGs), which address fundamental economic, social, and environmental requirements for sustainable progress. The role of technology in advancing sustainable development is of paramount importance and cannot be overstated. The United Nations recognized its importance and has mandated countries to develop their technology infrastructure as a tool to achieve the Sustainable Development Goals. Its use in in most countries of the world has created an effective and efficient means to achieve an all-round development affecting the state, its people and its environment. There is however, a greater need for low and middle income countries such as those that falls within the global south to embrace the use of technology by investing in its development across the country (Adebayo, 2018).

Technology contributes to strengthening national capacities for identified development challenges in areas such as environments, infrastructure, agriculture production, health, green technology, renewable energy and land degradation. In every sense, it holds the key to the accelerated growth and development of our cities by accelerating man-hour productive time. One of the thematic areas in the sustainable development goals that is closely linked to this chapter is goal 11 which emphasizes on making cities and human settlement inclusive, safe, resilient and sustainable. This is no exception to the achievement of smart cities with the impact of technology. As per Takase (2018), more than 50% of the global population currently resides in cities and urban areas, a figure projected to exceed 70% by 2050, primarily driven by growth in developing nations. To attain sustainable development goals, the integration of a reliable and affordable transportation system with emerging technologies is imperative for the realization of smart cities. Transport and mobility play a pivotal role in providing safe, affordable, accessible, and sustainable transportation systems for all, enhancing road safety by expanding public transport networks, with particular consideration for vulnerable groups such as women, children, persons with disabilities, and older individuals. Sustainable transport and mobility are integral to advancing the objectives outlined in the 2030 agenda for sustainable development and achieving the 17 Sustainable Development Goals (SDGs), as highlighted in the Global Mobility Report (2017). Sustainable transport directly contributes to the attainment of seven SDGs (1, 3, 8, 9, 10, 11, 13).

The Sustainable Mobility for All (SuM4All): a global multi-stakeholders partnership with a goal to make the mobility equitable, efficient, safe and green (clean) has the following objectives

- Safety: Improve the safety of mobility across all modes of transport by avoiding fatalities, injuries, and crashes from transport mishaps across all modes of transport, thus averting public health risks, and social and economic losses associated with unsafe mobility.
- Universal Access: Ensure that everyone (the elderly, children, women, families, and people with disabilities) has access to the transport needed, and "no one is left behind" to take advantage of social, economic, and environmental benefits for current and future generations. Equity and inclusiveness are the core of the global mobility objectives.

- Efficiency: Seek to ensure that transport demand is met effectively, at the least possible cost. Efficiency cuts across multiple aspects – the optimization of resources (i.e., energy, technology, space, institutions, and regulations) to generate an efficient transport system or network.
- Green Mobility: Aim to address climate change through mitigation and adaptation, and to reduce both air and noise pollution.

Based on the preceding considerations, it is evident that traditional mobility solutions are inadequate for addressing future transportation needs. Substantial improvements are imperative across several domains, including policy, planning, infrastructure, technology, and financing. One essential enhancement entails the adoption of integrated transport systems, facilitating seamless movement between different points and addressing the last mile connectivity challenge. Additionally, vehicles and transport infrastructure should be integrated into an intelligent network to enhance city safety, efficiency, and traffic flow. This will not help in addressing the problem of transportation but also, help in achieving a smart city which is inclusive, safe and innovative.

4.1.2 The Role of Technology in Sustainable Mobility

Transportation plays a crucial role in a country's economic development, necessitating the establishment of an efficient transport system (Salim, 2003; Lingaitiene, 2006). Achieving sustainable mobility is essential for the efficient, safe, secure, affordable, and environmentally friendly movement of people and goods. To attain the connectivity and efficiency goals of global trade, sustainable mobility and smart connectivity solutions are vital to sustain future growth (United Nations, 2021). Sustainable mobility is an integral part of smart sustainable cities, facilitating the movement of people and goods. It intersects with sustainable cities through innovative digital transport technologies (see Fig. 4.1), empowering city dwellers to make environmentally friendly choices and promoting sustainable economic growth while enhancing service delivery in cities. The rise of digital-based technologies, fueled by abundant data, has the potential to foster adaptive models in infrastructure, knowledge sharing and management, and collaborative relationships.

Among various transportation modes, private cars dominate, accounting for three-quarters of all passenger mobility. Over the past two decades, four competing technology solutions have emerged to decarbonize these light-duty motor vehicles: ultra-high efficiency internal combustion engines, biofuels, hydrogen, and electric vehicles (EVs). The use of electricity and hydrogen holds promise for decarbonizing the transport energy system, but their effectiveness relies on the utilization of low or zero-carbon energy sources. It is also important to consider the lifecycle environmental and health impacts of battery production and disposal for electric vehicles, especially as the global sales of EVs have been growing steadily (United Nations, 2021). Heavy-duty road transport, comprising trucks and buses, currently

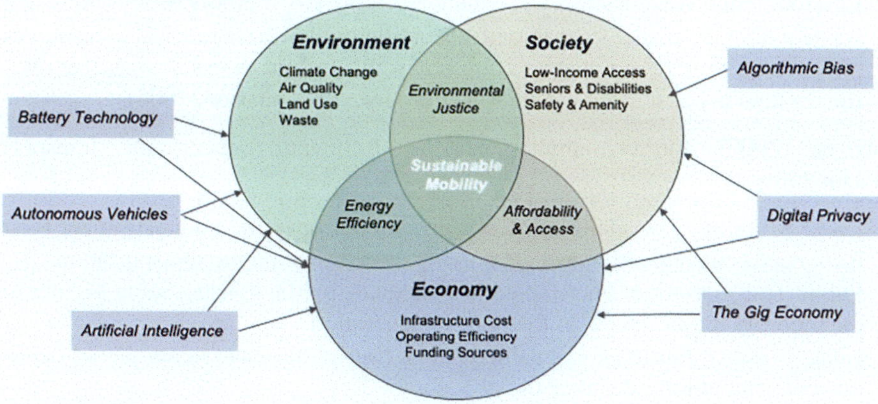

Fig. 4.1 Impact of technology on sustainable mobility. (Source: Borade et al., 2020)

contributes only 3% of global greenhouse gas (GHG) emissions. However, emissions from this sector are increasing by over 2% annually due to growing transport volumes, with trucks accounting for the majority of this growth. Addressing emissions from heavy-duty vehicles and road freight transport requires distinct solutions compared to light-duty vehicles, considering the challenges associated with developing battery and charger technology for long distances and heavy payloads. Although these technologies are at various stages of development, their cost remains significantly higher than that of alternative diesel vehicles.

While the future of smart connectivity holds promise, it is not without challenges. Barriers to technological deployment are evident in infrastructure limitations, knowledge management, and collaboration dimensions. Overcoming these challenges will be crucial for the successful implementation of smart connectivity in transportation systems.

4.1.3 Transport Sustainability

Transport sustainability encompasses various aspects that ensure the needs of individuals, companies, and society are met while prioritizing safety, human and ecosystem health, and intergenerational equity (European Commission, 2004). It is characterized by several key elements:

1. Accessibility and Development: A sustainable transport system provides safe and reliable access to basic necessities for individuals, businesses, and society. It promotes fairness, supports a competitive economy, and fosters balanced regional development.
2. Affordability and Efficiency: The system operates fairly and efficiently, offering a range of transportation modes and ensuring affordability for users. It contributes to a competitive economy and facilitates regional development.

3. Environmental Responsibility: Sustainable transport minimizes emissions and waste within the planet's capacity to absorb them. It utilizes renewable resources at a sustainable rate and manages non-renewable resources in a manner that minimizes land use and noise generation (European Commission, 2004).

Bayliss (2000) further emphasizes the following aspects of transport sustainability:

- Local Mobility: The majority of citizens should have the ability to meet their local needs conveniently through walking, cycling, or public transportation.
- Public Transportation: There should be adequate public transportation to cater to the mobility requirements of the present generation.
- Long-Distance Journeys: IT-enabled public transportation systems should facilitate convenient long-distance travel.
- Equity: No environmental or economic exceptions should favor private vehicle travel over sustainable modes of transportation.
- Efficient Road System: The capacity and management of the road network should align with the demand for highway capacity, taking into account full social cost pricing (Bayliss, 2000).

Assessing transport sustainability involves considering standards that either allow the substitution of natural capital with human capital or adhere to strong standards that reject such substitutions (Litman, 2008).

4.1.4 Information and Communication Technology (ICT) and Sustainable Transportation

Sustainable transport is a critical goal that requires the rapid deployment of innovative technologies to be achieved effectively. The transportation sector has witnessed the emergence of environmentally friendly fuels, advanced engines, autonomous vehicles, and intelligent transport systems (ITS) as key components of the transport innovation landscape (Vijayakumar & Mehendiratta, 2011). These advancements have been made possible by the significant role played by information and communication technology (ICT) in the transportation sector. ICT has facilitated the development of various applications aimed at supporting traffic congestion control, transport logistics, and transport infrastructure management (Vijayakumar & Mehendiratta, 2011). Additionally, ICT has expanded into the realm of information transportation, where it has the potential to enhance the efficiency of transport networks and mitigate their negative externalities, such as congestion and reduced network quality. However, the actual impact of ICT on the transport sector and sustainable development is still uncertain.

In developed cities like Stockholm, ICT has been instrumental in the development of sensing capabilities for vehicles and traffic monitoring systems (Weisers, 2008). This has enabled the detection of anomalies and hazards in the environment, resulting in improved traffic flow, cost savings, and a cleaner environment. To achieve sustainable transport planning, it is crucial to foster the widespread adoption and further development of ICT that promotes sustainable transport behavior (Weisers, 2008). Understanding the dynamic nature of technology within the transportation field is essential for successful sustainable transport initiatives.

4.1.5 Information and Communication Technologies that Supports Sustainable Mobility

To achieve a more sustainable transportation system, the importance of Information and Communication Technology (ICT) devices, particularly mobile phones, in traffic management and route assignment cannot be overstated. The widespread adoption of mobile technology has rapidly permeated various social groups, with global mobile phone penetration now exceeding 50% (Palmgren & Zapico, 2008). Since its inception as prototype equipment developed by Ericsson and Televerket in the 1950s, mobile phone technology has evolved from first-generation systems to the current second-generation technology known as GSM (Global System for Mobile Communication). This advancement has introduced services such as SMS, WAP, GPRS, and MMS, which have revolutionized communication by enabling internet connectivity, multimedia content sharing, and data transmission (Palmgren & Zapico, 2008).

Mobile phones serve as valuable tools for commuters as they facilitate pre-trip planning and provide real-time traffic information, effectively transforming them into multifunctional computer-like communication devices (Weisers, 2008). Offering a comprehensive view, mobile phones grant commuters primary access to communication technology while also playing a significant role in accessing value-added services like the internet. With the support of various applications, mobile phones empower commuters to easily and conveniently navigate their intended destinations. According to Aderibigbe (2021), telecommunication, particularly through the use of mobile phones, has significantly enhanced the majority of trips taken by road users, particularly in terms of time and space management. Consequently, telecommunication facilities, including mobile phones, can assist commuters in better planning their journeys to minimize traffic-related issues, such as trip cancellations or postponements.

4.1.5.1 Urban Computing

The urban environment can be categorized into three main layers: the natural layer, the constructed layer, and the virtual layer (Vijayakumar & Mehendiratta, 2011). The natural layer refers to the existing elements that are not manmade. The constructed layer encompasses human-made structures such as roads, bridges, and cars.

The virtual layer represents the integration of technology throughout the city, both visible and embedded. This includes personal computers, mobile phones, and various invisible technologies strategically placed in the city to support different activities. An example of urban computing is the implementation of traffic algorithms to manage and control traffic at road intersections. The concept of urban computing combines urbanity with ubiquitous computing, where computers become an integral part of the urban fabric.

4.1.5.2 Ubiquitous Technologies

Ubiquitous computing, developed by Mark Weiser, represents the next generation of computing that moves away from the desktop model (Weisers, 2008). It integrates information processing into everyday objects and activities. Unlike the traditional desktop paradigm, ubiquitous computing involves simultaneous engagement with multiple computational devices and systems, often within the context of ordinary activities. The advantage of ubiquitous technologies is that they seamlessly integrate technology into the user's environment, providing real-time information and enhancing user experiences. Applications of ubiquitous computing range from GPS-based car navigation systems to handheld tour guide systems that adjust content based on the user's physical location (Abowd & Mynatt, 1999).

4.1.5.3 Dedicated Short Range Communication (DSRC) Charging Using Transponders and Gantries

DSRC is an advanced pricing technology commonly used for congestion control on toll facilities. The On-Board Unit (OBU), also known as tags or transponders, connects to equipment mounted on gantries at checkpoints (Federal Highway Administration, 2006). When a vehicle equipped with a tag enters or exits the defined area, the system charges the account or records the vehicle's entrance. Enforcement is carried out using roadside enforcement cameras or a combination of Automatic License Plate Recognition (ALPR) technology. Infrared communication was initially used but proved inadequate for high-speed applications. The current implementation of DSRC utilizes microwave communications operating at a frequency of 5.8 GHz, following the European CEN-278 standard (Federal Highway Administration, 2006). This system relies on robust and secure OBUs with an average battery life of 5 years. Expansion of the DSRC system to additional tolls or checkpoints is easily achievable, providing cost-effective coverage over wider areas. The street-based DSRC system requires various components such as transceiver-equipped poles or gantries, illumination devices, vehicle classification devices, independent verification devices, ALPR cameras, and roadside cabinet controls. The transceiver unit is usually separate from the cameras to enable vehicle detection in the pricing zone, although combining functions at a single location is possible through accessible technologies (Lindsey & Palma, 2009) (Fig. 4.2).

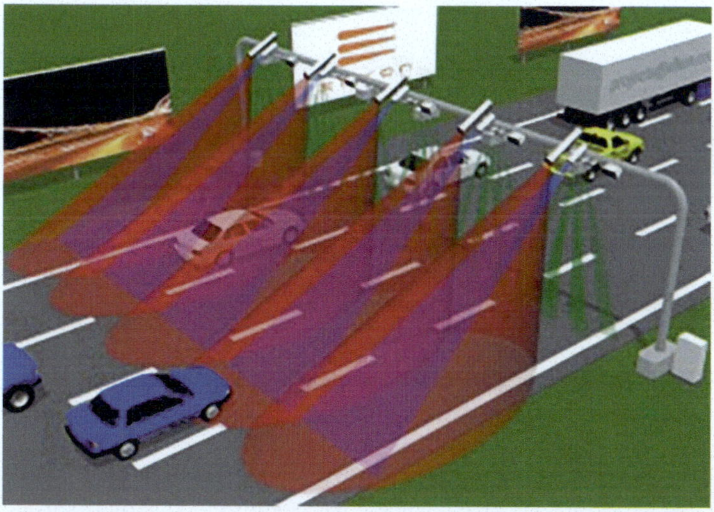

Fig. 4.2 A vehicle being detected while entering the congested area by a DSRC system (Bayliss, 2000)

4.1.5.4 Automatic License Plate Recognition (ALPR) Technology

Automatic license plate recognition (ALPR) technology is commonly utilized in electronic congestion pricing systems as a complementary enforcement method alongside DSRC systems or VPS (Ezell, 2010). This technology captures images of vehicle license plates within a designated area and employs character recognition software to identify the vehicle and its owner. To enhance identification rates, modern electronic pricing systems often incorporate front and rear cameras to capture images from both sides of the vehicle. The ALPR system requires pole or gantry mounted cameras and an illumination device, typically installed on roadsides or streets (see Fig. 4.3). In addition to congestion pricing, ALPR technology can be applied for tracking vehicle locations to enhance safety and support transport planning initiatives.

4.1.5.5 Vehicle Positioning System (VPS) Technology

Vehicle Positioning System (VPS) technology, such as the Galileo Global Navigation Satellite System (GLONASS) and GPS, is aimed at mitigating traffic congestion (IBM, 2007). VPS utilizes satellite location systems, predominantly GPS, to determine the vehicle's position and charge appropriate fees for accessing traffic-controlled areas. These systems offer flexibility to authorities, allowing charges to be adjusted based on travel time or specified entry zones, influencing travel behavior and transportation choices. VPS-based congestion pricing provides an effective

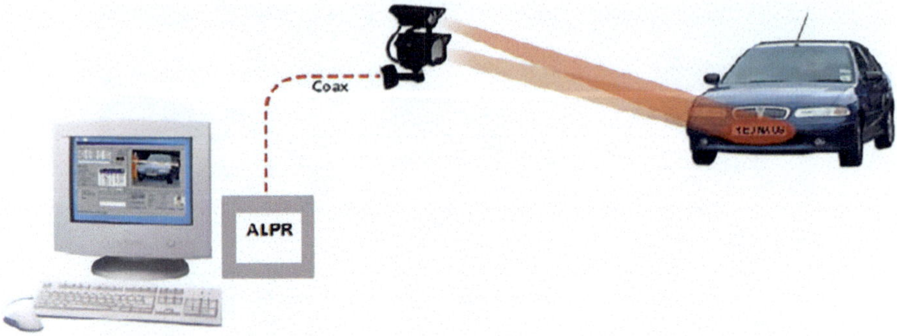

Fig. 4.3 A simplified form of ALPR system. (Adapted from Bayliss, 2000)

approach to balance diverse requirements. The VPS-based pricing system operates through an In-Vehicle Unit (IVU) and a control center system. It enables efficient vehicle tracking, and the gathered information is transmitted to the control center for charge processing and enforcement procedures. VPS systems can be integrated with other technologies, such as wide area communication, Google Maps, and short-range GPRS, and combining them with DSRC systems, including 5.9-GHz technology, Wi-Fi, WiMAX, and 3G communication capabilities, enhances the popularity of vehicle positioning systems for congestion pricing (Ezell, 2010).

4.1.5.6 Vehicle Occupancy Detection Technologies

Vehicle Occupancy Detection Technologies are considered highly effective ICT tools for promoting sustainable transportation mobility and minimizing traffic congestion. These systems enable congestion charging based on vehicle occupancy. In some countries like the U.S., this approach is implemented to categorize travel through High Occupancy Vehicle (HOV) lanes, granting access to ride-sharers and reducing congestion charges for vehicles not falling under the "drive alone" category (Diamond, 2008). This strategy encourages commuters and car-poolers to utilize high occupancy vehicles, providing access to designated lanes and thereby alleviating traffic congestion. While existing systems can analyze Closed-Captioned Television Images to determine the number of travelers in a vehicle, ongoing technological advancements and collaborative efforts have improved the software's capability to detect human occupants. Through the integration of these ICT solutions in transportation, issues such as traffic congestion and accidents can be effectively addressed, fostering sustainable planning.

4.1.6 Active Travel and Sustainable Transport Planning

The International Energy Agency forecasts a substantial 120% rise in greenhouse gas emissions from the transport sector between 2000 and 2050 (ITDP, 2015 as cited in Tønnesen et al., 2020). Consequently, there has been an increasing focus on reducing reliance on private vehicles and fostering walking and cycling as integral components of a sustainable transportation network (GCPH, 2022). Active and sustainable travel not only benefits population health but also contributes to environmental preservation. Therefore, public health leaders and politicians advocate for greater investment in sustainable, integrated transport to encourage active travel, create safer communities, mitigate climate change effects, and reduce air pollution. However, current trends show an increase in car usage, low levels of cycling, declining walking and bus use, and rising road traffic and vehicular emissions. These trends exacerbate inequalities as vulnerable groups, including children, the elderly, and those living in poverty, bear a disproportionate burden of negative impacts such as traffic casualties and air pollution.

Active travel, encompassing walking, cycling, and other non-motorized modes, plays a pivotal role in building accessible, safe, inclusive, and green cities. In Latin America and the Caribbean, walking and cycling constitute a significant proportion of total trips, particularly among low-income groups, offering sustainable and affordable transportation options for short and medium distances. However, for low-income individuals, the decision to use active transport modes, especially walking, is primarily driven by affordability rather than sustainability.

While the benefits of walking and cycling in terms of health, environment, and quality of life are widely acknowledged, these active transport modes often receive limited attention and investment from authorities in many cities of Latin America and the Caribbean (IDB, 2021). The lack of an enabling environment hinders their integration with other modes, especially public transport, impeding access to employment opportunities and essential services for low-income groups residing in peripheral areas. Connectivity infrastructure and services are essential for bridging these gaps. Research by Cavallo et al. (2020) indicates that people are willing to sacrifice motorized trips, including public transport, due to affordability, leading them to rely on non-motorized transport systems like walking or cycling. This suggests that low-income earners adopt active modes not solely for their health and environmental benefits but because of their economic circumstances, resulting in longer travel times and impacting their quality of life. For example, analysis of selected cities in Latin America and the Caribbean shows that the bottom quintile of income earners in half of the cities spend more than 25% of their monthly income on a basket of 60 trips (Rivas et al., 2019).

In Montevideo, Uruguay, Mauttone and Hernández (2017) found that 27% of individuals from low-income groups walk because they have no other means of transportation, compared to only 5% of high-income individuals. Socio-economic differences in cycling trips are less pronounced, with variations depending on urban commuting patterns and the size of the city. For instance, in Bogota, Colombia, individuals from low-income groups cycle more frequently than higher-income groups, with cycling trips representing 4.8% and 9.0% of all trips in the poorest and second-poorest income groups, respectively. These individuals often spend more time traveling longer distances to avoid public transport costs (Alcaldía de Bogotá, 2020).

4.1.7 Benefits of Active Travel

The role of active travel plays a significant role in both individual lives and the community as a whole. One of its key benefits is the reduction of harmful gaseous emissions into the environment, as urban areas are responsible for over 60% of global greenhouse gas (GHG) emissions, with approximately one-third of these emissions originating from urban transport (World Bank, 2021). The World Bank (2021) conducted a study in selected European cities, revealing that changing transportation habits to rely less on single occupancy vehicles, such as private cars that utilize fossil fuels, and adopting sustainable modes of transport like non-motorized options (walking/cycling) can lead to a reduction in carbon footprint. Moreover, the provision of pedestrian walkways in urban centers can significantly reduce air pollution. For instance, in Spain, city centers have been pedestrianized since 1999, resulting in a remarkable 70% decrease in carbon dioxide emissions (Burgen, 2018).

Beyond environmental benefits, active travel has been shown to improve cardiovascular health, reduce obesity, and mitigate various health conditions by engaging individuals in physical activities like walking and cycling. The World Health Organization (WHO) (2020) reports that around 3.2 million deaths annually are attributed to physical inactivity. Essel and Spadaro (2020) highlight that active travel can help address some of these deaths. Their study suggests that implementing sustainable mobility options, such as safe cycling and walking, could potentially save up to 5500 premature deaths through improved air quality and an additional 33,000 lives in Accra, Ghana, over a 35-year period by increasing physical activity levels.

In addition to the various advantages it offers, active travel also brings about economic development through investments in bicycle and pedestrian infrastructure, resulting in positive impacts on local retail and food service businesses. According to Volker and Handy (2021), the implementation of active mobility infrastructure leads to increased sales in retail and food sectors, job opportunities, and customer spending. Moreover, the presence of bicycle and pedestrian facilities attracts more visitors who choose to travel by foot or bike, further contributing to the economy. This economic boost has even led to the emergence of new industries like green tourism, as evidenced by the significant annual impacts in Europe (44 billion euros) and the United States (USD 96.7 billion) attributed to bicycle tourism (Adventure Cycling Association, 2017). Improving active mobility within cities and communities also brings about social benefits such as enhanced social cohesion, improved perceptions of security, and overall livability. Traveling alongside others on foot or by bike fosters a sense of fulfillment and positive experiences, as highlighted by McIlvenny (2015). The COVID-19 pandemic and subsequent lockdowns have significantly impacted people's physical and mental well-being. Active travel modes provide opportunities for social interactions among diverse community members, facilitating engagement with differences and promoting social cohesion. Ultimately, the availability of safe environments for active travel positively influences individuals' physical, mental, social, and economic well-being. Overall, active travel plays a pivotal role in the creation of inclusive, safe, resilient, and sustainable cities and human settlements.

4.1.8 Ways of Promoting Active/Non-motorized Transport

Non-motorized transportation (NMT), also referred to as active transportation and human-powered transportation, encompasses walking, bicycling, and various modes such as cycle rickshaws, skates, skateboards, push scooters, hand carts, and wheelchair travel (Witting & Godefrooij, 2009). These modes are particularly significant for short trips within urban areas, which constitute the majority of journeys (Witting & Godefrooij, 2009). Promoting NMT requires a comprehensive policy package that includes investments in infrastructure, awareness campaigns, smart urban planning, improved public transport, and disincentives for motorized private vehicle usage. To encourage active travel, it is essential to prioritize the provision of safe and convenient infrastructure that accommodates pedestrians and cyclists of all ages, genders, and abilities. Infrastructure planning and design, along with traffic enforcement measures, play a crucial role in protecting individuals from motor vehicle collisions (Litman, 2009). Key measures for promoting active travel include:

- Improvement of sidewalks, crosswalks, paths, bicycle lanes, and networks.
- Implementation of public bicycle systems, which offer automated bicycle rental for short urban trips.
- Development of pedestrian-oriented land use and building designs.
- Enhancement of road and path connectivity, including the creation of non-motorized shortcuts.
- Implementation of traffic calming measures, streetscape improvements, speed reductions, vehicle restrictions, and reallocation of road space.
- Provision of safety education, law enforcement, and encouragement programs.
- Establishment of bicycle parking facilities.
- Integration of bicycles into transit systems, such as providing racks on metros or buses.
- Addressing security concerns for pedestrians and cyclists.
- Implementation of congestion pricing.
- Implementation of vehicle parking policies and fuel taxes.

Additionally, to promote active travel, it is crucial to design pedestrian networks that are uninterrupted, interconnected, easily accessible, pleasant, and safe

(SLOCAT, 2021). Uninterrupted and interconnected networks should offer multiple well-connected pathways to essential destinations. Accessible and pleasant streets should ensure inclusivity and ample space for pedestrians to walk comfortably. Safety measures should encompass well-illuminated, unobstructed, and easily accessible areas, along with clearly marked and visible crossings at intersections. Overall, non-motorized transportation plays a vital role in promoting human health, environmental improvement, and sustainability.

4.1.9 Public Transport and Sustainable Transport Planning

Sustainable transport is recognized as a crucial pathway to achieving universal access, enhanced safety, reduced environmental and climate impact, improved resilience, and greater efficiency (United Nations, 2021). It plays a vital role in realizing the 2030 Agenda for Sustainable Development by contributing to goals such as poverty eradication, reduced inequality, women's empowerment, and climate change mitigation. The development of an effective transportation network is essential for the economic and social progress of cities. However, as cities have expanded, transportation networks have presented challenges to achieving sustainability. These challenges include congested roads resulting from heavy reliance on automobiles, leading to emissions and social costs like accidents (Moavenzadeh et al., 2002a, b). While transportation networks drive urban development, they also create economic, social, and environmental impacts (Banister, 2005).

Investing in public transit is widely regarded as a measure to address transportation problems, particularly traffic congestion, and reduce the negative impacts of transportation networks on society and the environment (Banister, 2005). Public transit is seen as a key driver of sustainability in the transport sector and a means to mitigate the sustainability impacts of auto-dependence (Miller et al., 2016). It offers several benefits, including energy efficiency, reduced emissions, improved access, increased economic efficiency, and economic activity (Schiller et al., 2010). Public transit's energy efficiency positions it as a sustainable alternative to private automobiles, thereby reducing dependence on cars and associated impacts (Schiller et al., 2010). The space efficiency and social benefits of transit contribute to sustainability across multiple criteria, making it a pivotal factor in reducing auto travel and dependence in cities (Schiller et al., 2010). To achieve sustainable development and urbanization goals through a shift to transit, a reallocation of public space previously allocated for automobile travel is necessary (Banister, 2008).

Transit leverage, as proposed by Newman and Kenworthy (1999), emphasizes the benefits of substituting car trips with transit trips, resulting in a reduction in passenger kilometers traveled within a network. Good transit options influence businesses and individuals to adjust their location behavior, while transit users combine trips and often embrace active travel modes like walking or cycling to access stations or stops, thereby promoting sustainability (Newman & Kenworthy, 1999). Investing in public transport not only benefits individuals but also contributes to economic growth. For instance, the Canadian Urban Transit Association estimates that public transit provides a $10 billion benefit to the Canadian economy annually (Canadian Urban Transit Association, 2010). Therefore, it is essential to prioritize investment in the public transport system to fully harness its advantages (Banister, 2008).

4.1.10 Integrated Transport Planning for Sustainable Development

Transportation plays a crucial role in the political, economic, and social development of both rural and urban societies. It serves as the primary means of connecting different parts of a society (Aderamo & Magaji, 2010; Dosunmu & Mogbojuri, 2019). Over the years, there has been a significant increase in traffic volume globally, driven by the growing demand for mobility resulting from increased car ownership in both developed and developing countries (Andrzej, 2002). However, this rise in traffic has led to major traffic congestion issues in cities worldwide, negatively impacting their attractiveness through congestion, noise pollution, and air pollution from exhaust fumes. To address these challenges, sustainable urban transport systems require alternatives to private car usage that are convenient, flexible, and cost-effective (Andrzej, 2002).

Merely expanding existing transportation infrastructure may not be an effective solution to the ever-growing traffic problems. Instead, there is a need to adopt an integrated transportation approach. Integrated Transportation Management Systems aim to integrate all modes of transportation and roads into a cohesive system, with researchers focusing on the benefits that can be achieved through such integration (Bibiana, 2013). An integrated transport system seeks to establish a seamless network of interconnected and complementary transportation modes, enabling each mode to fulfill its potential in a collaborative manner. Through shared infrastructure, fare structures, and standard approval systems, integrated transport systems allow public transit users to utilize different modes of transportation in a coordinated manner (Bibiana, 2013).

In an integrated transport system, three main components are identified (Bibiana, 2013):

1. Suppliers of public passenger transport: They play a key role in interconnecting their systems, considering transportation, economic, hierarchical, and levy aspects.
2. Users of public passenger transport: They are the decision-makers, driving the success of the transport system through their behavior.
3. Authorities: They establish the legislative structures that govern the operation of the integrated transport framework, transitioning requirements from residents to operators.

For operators involved in the integrated transport system, key areas of focus include:

1. Modal integration: This aims to minimize journey time from origin to destination, reducing passenger time loss.
2. Organizational-economic integration: It involves managing and allocating financial flows between the involved parties, including revenue from fares and subsidy streams to cover costs not covered by revenues.
3. Tariff integration: This entails creating a unified fare system that is easily understandable to passengers. The integrated transport system divides the area into regular or irregular zones, and fares are determined based on the tax structure, providing ticket options for regular and occasional passengers.

By adopting an integrated transport system, cities can address the challenges associated with traffic congestion and enhance the efficiency of their transportation networks, contributing to sustainable development and improved quality of life.

4.1.11 Methods of Integrating Transport Systems

1. **Design interchange stations to provide secure, uncongested conditions for transfer by the shortest routes possible**: One method to integrate transport systems is by designing interchange stations that provide secure and uncongested conditions for transfers, prioritizing the shortest routes possible. It is important to ensure that commercial outlets in stations are carefully designed to not hinder access to public transport services. An example of this can be observed in the varying quality of pedestrian access to cross rail interchange stations in London (Smith et al., 2018).
2. **Provide adequate pedestrian/cycling walkways such as bike parking areas at stations and stops**: To achieve full integration of the transport system, it is crucial to provide adequate pedestrian and cycling walkways, including bike parking areas at stations and stops. This accommodates non-motorized transport users and facilitates cycling for the last mile. However, it is important to address the issue of undisciplined bike parking that can block access to stations. As cycling becomes more popular, the need for off-street parking facilities and reallocation of space arises (Gumbo & Moyo, 2020).
3. **Integrate ticketing and information systems as well as the physical transport infrastructure**: Integrating ticketing and information systems, along with the physical transport infrastructure, is also crucial. Integrated ticketing systems allow for the use of a single physical or electronic ticket to pay for multiple modes of transport and even non-transport goods and services. Countries like Korea have set examples with their integrated smart travel payment systems, utilizing pre-paid smart cards and bank cards with integrated travel functionality (Gumbo & Moyo, 2020; OECD, 2018). Additionally, the introduction of smart phone payment systems for bike-share systems has expedited the process for occasional users.

4. **Establish integrated urban transport plans in consultation with stakeholders and the public**: Establishing integrated urban transport plans through consultation with stakeholders and the public is essential. This process ensures consensus among the community and delivers sustainable outcomes. It is important to consider planning and investment horizons that extend beyond political electoral cycles, as stakeholder consensus provides a long-term strategic perspective (Brown & Green, 2019).
5. **Integrated fares**: Integrated fares play a vital role in promoting multi-modal transport usage. By offering fare options that are independent of the mode or number of services used for a journey, passengers are encouraged to utilize different modes of transport. The concept of "mobility as a service" demonstrates the bundling of payments for various transport options into a single app platform, excluding private cars (Aderibigbe, 2022).
6. **Integrated service information**: Integrated service information ensures that travelers can obtain information from a single source in a consistent format, regardless of the mode of transport they are using. This simplifies the journey planning process and avoids the need to consult different sources for each component of a multi-modal journey (Hall et al., 2019).

4.1.12 The Roles of Integrated Transport Systems in Sustainable Planning and Development

Integrated transport systems contribute to sustainable planning and development in various ways:

1. **Integrate land use and transport planning**: They facilitate the integration of land use and transport planning by promoting efficient land use and traffic policy enforcement, which supports compact community development and land use intensification while preserving natural habitats (Hall et al., 2019).
2. **Improve public transport**: Improved public transport plays a crucial role in sustainable transport development. By introducing a modern, integrated, high-quality, affordable, and customer-focused public transport system, the quality of life for individuals is enhanced. Furthermore, it enhances the safety, security, resilience, reliability, and efficiency of the public transport network (Litman, 2019).
3. **Providing universally accessible transport**: Integrated transport systems aim to provide universally accessible transport, catering to all stakeholder groups, including persons with disabilities, across different regions and provinces (Cui et al., 2022).
4. **Promoting the use of new and more efficient technologies**: These systems promote the use of new and efficient technologies, intelligent transport solutions, and energy-efficient options for infrastructure and systems (Loo et al., 2020).

5. **Promote safety and well-being**: They contribute to safety and well-being by eliminating traffic-related fatalities, addressing personal safety and security concerns, and ensuring people of all ages and statuses canarrive at their destinations safely. This fosters inclusive and equitable transport systems within the community (Wegener et al., 2019).
6. **Ensure social inclusion and accessibility**: Integrated transport systems ensure social inclusion and accessibility by providing access to vibrant public open spaces that encourage walking, cycling, social interaction, and active lifestyles. This improves accessibility and connectivity, particularly for marginalized communities (Aderibigbe, 2022).
7. **Promote freight integration**: Promoting freight integration is another role of integrated transport systems. By coordinating and integrating freight transport operations across various modes, a more balanced distribution of freight service delivery can be achieved. This reduces over-reliance on a single mode, such as road transport, and enhances overall efficiency (Gössling, 2020).
8. **Protect the environment and overall liveability of the city**: Protecting the environment and enhancing the overall liveability of cities is a significant benefit of integrated transport systems. By promoting active travel, such as walking and cycling, and encouraging public transport usage, these systems help reduce harmful emissions and mitigate the impact of transportation on the environment. Additionally, the adoption of low-carbon and renewable energy sources further contributes to environmental protection and the preservation of the ozone layer (Zheng et al., 2021).
9. **Support rural development**: Integrated transport systems also support rural development by creating nodes and linkages to major cities and economic centers. This connectivity improves access to opportunities for rural communities and contributes to their economic growth and well-being (Givoni & Perl, 2020).

Conclusively, integrated transport systems play a crucial role in achieving sustainable growth and overall development in society. By addressing issues of accessibility, safety, environmental impact, and social inclusion, these systems create safer, more comfortable, and livable communities. Through the integration of land use and transport planning, the improvement of public transport, the promotion of new technologies, and the consideration of stakeholder perspectives, integrated transport systems pave the way for smart and sustainable mobility.

4.1.13 Challenges to Achieving Sustainable Transport

There is a pressing need for a comprehensive analysis of the challenges and prospects related to sustainable transport in various countries. To address this, global policies and initiatives are required to shape the development of transportation in both developed and developing nations. However, several challenges hinder the achievement of a seamless and sustainable transport system.

Demographic Change: One major challenge is demographic change. The global population is expected to reach 8.5 billion by 2030, with varying growth rates across regions. Sub-Saharan Africa is experiencing rapid population growth, while Europe is projected to face a decline in population between 2015 and 2030 (United Nations, 2021). Moreover, countries in Europe and Asia are witnessing an aging population, with individuals aged 65 and over surpassing the number of children under five in 2018. By 2050, it is estimated that there will be over twice as many people over 65 as children under five, and the number of elderly individuals will exceed the number of adolescents and youth aged 15–24 years. This demographic trend is expected to strain sustainable transport services. With larger populations, transportation systems must handle increased demand, while finding ways to mitigate global threats such as climate change and biodiversity loss. Additionally, countries with aging populations will require inclusive and accessible transport services, resulting in a greater need for equitable and affordable transport solutions.

Urbanisation: Urbanization is another factor impeding sustainable transport planning. Currently, nearly 4 billion people reside in cities, and this number is projected to exceed 6 billion by 2050. By that time, approximately 70% of the global population will be living in cities, generating 85% of the world's economic output. The rapid increase in urban populations with high population densities poses challenges due to heavy reliance on fossil fuel-powered vehicles. This reliance contributes to issues such as air and noise pollution, traffic congestion, and road safety hazards. The World Health Organization estimates that around seven million people die annually from inhaling polluted air, leading to various illnesses. Transport significantly contributes to these impacts, with vehicle tailpipe emissions alone responsible for approximately 385,000 premature deaths worldwide in 2015. Diesel vehicle exhaust emissions accounted for nearly half of these impacts. Urbanization thus poses a significant threat to achieving a sustainable transport future.

Globalization: Globalization is also a factor that affects sustainable transportation. A globalized economy relies on international trade, which creates a demand for high-volume freight transport and complex global supply chains. Additionally, tourism relies on passenger transport across all modes, and migration is facilitated by various transport routes and modes. International trade and passenger travel are key indicators of an interconnected world. For example, merchandise trade was valued at approximately $20 trillion in 2018, having grown by 26% over the previous decade, keeping pace with global real GDP growth (United Nations, 2021). These trends are expected to exert pressure on transportation systems, particularly with increased trade and tourism development.

Digitalization: On the other hand, digitalization presents prospects for sustainable transportation. In the era of globalization, digitalization has emerged as a defining characteristic of the 2020s. It involves the increased use of various digital technologies, including the Internet of Things, advanced robotics, industrial biotechnology, 3D printing, new materials, and nanotechnology. These technologies are already in use and are being adopted at an accelerating pace worldwide. The deployment of artificial intelligence (AI) further enables and accelerates these trends, automating cognitive processes and improving the detection, prediction, and response to people's behaviors and customer needs, including in the transport sector.

The advent of digitalization holds immense transformative potential for global value chains, revolutionizing their management and forecasting capabilities. This has wide-ranging implications for productivity, employment, skill development, income distribution, trade, well-being, and environmental sustainability. Additionally, digitalization empowers individuals to access a plethora of online services, experiences, and employment opportunities right from the convenience of their homes. As a result, there is a potential reduction in the need for travel, encompassing commuting and business trips. Consequently, digitalization presents opportunities to make the transportation sector more ecologically friendly, provided it does not lead to increased leisure travel and last-mile freight transportation.

In conclusion, to ensure sustainable transport planning, substantial investments are necessary to enhance the physical robustness of infrastructure through the adoption of technologies and other measures. These investments should prioritize areas such as public transport and active travel to ensure that transport systems remain functional, recover quickly, and incur minimal costs (Smith et al., 2020). By addressing challenges related to demographic change, urbanization, globalization, and leveraging the prospects of digitalization, countries can work towards achieving sustainable transport systems that promote environmental conservation, accessibility, and equitable mobility.

References

A Primer: Federal Highway Administration. (2006). *Congestion pricing*. Federal Highway Administration.

Abowd, G. D., & Mynatt, E. D. (1999). *Ubiquitous computing: Past, present and future*. GVU Center and College of Computing.

Abubakar, A. B. (2013a). Gender studies and sustainable development in Nigeria. *Journal of Educational and Social Research, 3*(10) Retrieved from www.mcser.org/journal/index.php/jesr/article/view/2339

Abubakar, B. (2013b). Environmental sustainability and sustainable development: The case of Nigeria. *Journal of Environmental Protection, 4*(11), 1367–1378.

Adebayo, T. J. (2018). *The role of technology in achieving the SDGs 2030*. https://www.academia.edu/38900809/THE_ROLE_OF_TECHNOLOGY_IN_ACHIEVING_THE_SDGs_2030. Assessed 12 Jan 2024.

Aderamo, A., & Magaji, S. (2010). *Rural transportation and the distribution of public facilities in Nigeria: A case of edu Local Government Area of Kwara State*.

Aderibigbe, O. (2021). Are telecommunications and travel substitutes or complements: An empirical analysis of a developing city in Nigeria. *Transport Findings*. https://doi.org/10.32866/001c.30742

Aderibigbe, O. (2022). Adoption of Integrated Transport System for Sustainable Transport Planning and Development. Conference Proceedings for International Symposium on

Inclusive-Cities: Achieving Inclusive Cities Through A Multidisciplinary Approach, 2021 28–30 June. *Journal of Inclusive cities and Built environment, 2*(1), 41–44. https://doi.org/10.54030/2788-564X/2022/cp1v2a9

Adventure Cycling Association. (2017). *Bicycle Tourism 101*. https://www.adventurecycling.org/advocacy/building-bike-tourism/bicycle-tourism-101/

Alcaldía de Bogotá, D. C. (2020). Encuesta de Movilidad 2019: Indicadores Preliminares. Secretaría de Movilidad, Bogotá D.C. https://www.movilidadbogota.gov.co/web/sites/default/files/Paginas/22-04-2020/20191216_presentacion_encuesta_v2.pdf

Andrzej, A. (2002). *Intelligent integrated transport systems*. 565–570.

Ayeni, A. (2010a). Sustainable development through international cooperation in research and education. In L. Popoola & O. Olorunnisola (Eds.), *Contemporary issues in sustainable development: Lessons for, and challenges to Nigeria*. The Postgraduate School University of Ibadan.

Ayeni, D. (2010b). Sustainable development: Issues and challenges. *Journal of Sustainable Development, 3*(4), 36–44.

Banister, D. (2005). *Unsustainable transport: City transport in the new century*. Routledge.

Banister, D. (2008). Cities, mobility, and climate change. *Journal of Industrial Ecology*, 7–10. https://doi.org/10.1162/jie.2007.1271

Barbier, E. B. (1989). *Economics, natural resources, security and development: Conventional and alternative views*. East Asia Publishers.

Bayliss, D. (2000). *Urban development and its implications for mobility*. International Union of Public Transport (UITP) Mexico Conference.

Bibiana, P. (2013). Key success factors of integrated transport. In *The 13th international conference 'Reliability and statistics in transportation and communication'* (pp. 83–90).

Black, W. (2010). *Sustainable transportation: Problems and solutions*. Guildford.

Borade, S., Cohen, K., Davis, C., & Stockman, J. (2020). *Sustainability mobility in the United States: Challenges and opportunities from new technologies*.

Brown, A. H., & Green, T. D. (2019). The Essentials of Instructional Design Connecting Fundamental Principles with Process and Practice. https://doi.org/10.4324/9780429439698

Burgen, S. (2018, September 18). For me, this is Paradise: Life in the Spanish City that Banned Cars. *The Guardian*. https://www.theguardian.com/cities/2018/sep/18/paradise-life-spanish-city-banned-cars-pontevedra

Canadian Urban Transit Association. (2010). *Measuring success: The economic impact of transit investment in Canada*. Retrieved from Canadian Urban Transit Association: http://www.cuta-actu.ca/en/publicationsandresearch/resources/Issue_Paper_35E.pdf

Cavallo, E. A., Powell, A., & Serebrisky, T. (2020). *From structures to services: The path to better infrastructure in Latin America and the Caribbean*. Inter-American Development Bank.

Cui, Y., Chen, X., Chen, X., & Zhang, C. (2022). Competition, Integration, or Complementation? Exploring Dock-Based Bike-Sharing in New York City. *The Professional Geographer, 75*(1), 65–75. https://doi.org/10.1080/00330124.2022.2081224

Diamond, D. (2008). Incentives for high occupancy vehicle (HOV) lane incentives for hybrids in Virginia. *Journal of Public Transportation, 11*(4), 55–62.

Dosunmu, V. A., & Mogbojuri, O. (2019). Perspective of integrated transport systems for sustainable development in an emerging economy. *International Journal for Communications and Networks, 2*(8) ISSN:(2637-6571).

Drexhage, J., & Murphy, D. (2010). *Sustainable development: From Brundtland to Rio 2012*. Background Paper for the World Economic and Social Survey 2011. Retrieved from http://www.un.org/wcm/webdav/site/climatechange/shared/gsp/docs/GSP1-6_Background on Sustainable Development.

Essel, D., & Spadaro, J. V. (2020). *Health and economic impacts of transport interventions in Accra, Ghana*. World Health Organization. https://www.who.int/publications-detail-redirect/9789240017306

European Commission. (2004). *Towards a thematic strategy on the urban environment*. Commissions of the European Community.

Ezell, S. (2010). *Intelligent transport system*. The Information Technology & Innovation Foundation.

Fifth EC Environmental Action Programme. (1993). *Towards sustainability*. Retrieved from http://ec.europa.eu/environment/archives/action-programme/5th.htm.

Givoni, M., & Perl, A. (2020). Rethinking Transport Infrastructure Planning to Extend Its Value over Time. *Journal of Planning Education and Research, 40*(1), 82–91. https://doi.org/10.1177/0739456X17741196

Glasgow Center for Population Health. (2022). *Informing and supporting action to improve health and tackle inequality*. https://www.gcph.co.uk/transport_and_travel/active_sustainable_travel. Accessed 2 Dec 2022.

Gössling, S. (2020). Why cities need to take road space from cars – and how this could be done. *Journal of Urban Design, 25*(4), 443–448, https://doi.org/10.1080/13574809.2020.1727318

Gumbo, T., & Moyo, T. (2020). Exploring the interoperability of public transport systems for sustainable mobility in developing cities: Lessons from Johannesburg Metropolitan City, South Africa. *Sustainability, 12*(5875), 1–16. https://doi.org/10.3390/su12155875

Hall, P., Hall, P., & Tewdwr-Jones, M. (2019). *Urban and Regional Planning* (6th ed.). Routledge. https://doi.org/10.4324/9781351261883

IBM. (2007). *Driving intelligent transport systems*. IBM.

Inter-American Development Bank (IDB). (2021). *The role of active transport modes in enhancing the mobility of low-income people in Latin America and the Caribbean*.

Lindsey, R., & Palma, A. D. (2009). *Traffic congestion pricing methods and technologies*. Departement d'economie.

Lingaitiene, O. (2006). Application of a general model of multimodal transportation in logistics. *Transport and Telecommunication, 7*(1): 92–96.

Litman, T. (2008). *Sustainable transportation indicators: A recommended research program for developing sustainable transportation indicators and data*. A paper presented at 2009 Transportation Research Board annual meeting.

Litman. (2009, September 24). *Quantifying the benefits of nonmotorized transportation for achieving mobility management objectives*. Victoria Transport Policy Institute.

Loo, B. P. Y., Tsoi, K. H., & Banister, D. (2020). Recent experiences and divergent pathways to transport decoupling. *Journal of Transport Geography, 88*, 102826. https://doi.org/10.1016/j.jtrangeo.2020.102826

Litman, T., (2019). *Understanding Transport Demands and Elasticities. How Prices and Other Factors Affect Travel Behavior*. Victoria Transport Policy Institute. Retrieved from. https://www.vtpi.org/elasticities.pdf

Low, N. (2003a). Dimensions of sustainability. In *Encyclopaedia of global environmental change* (Vol. 1, pp. 412–418). Wiley.

Low, N. (2003b). Is urban transport sustainable? In *In making urban transport sustainable* (pp. 1–22). New York, NY.

M3 Future. (2005). *Sustainable development, an introduction*. Retrieved from http://www.m3future.com/pressrel/ssd/sdintro.htm

Mauttone, A., & Hernández, D. (2017). *Encuesta de movilidad del área metropolitana de Montevideo. Principales resultados e indicadores*. CAF, Intendencia de Montevideo, Intendencia de Canelones, Intendencia de San José, Ministerio de Transporte y Obras Públicas, Universidad de la República, PNUD Uruguay.

McIlvenny, P. (2015). The joy of biking together: Sharing everyday experiences of Vélomobility. *Mobilities, 10*(1), 55–82. https://www.tandfonline.com/doi/abs/10.1080/17450101.2013.844950

Miller, P., De Barros, A., Kattan, L., & Wirasinghe, S. C. (2016). Public transportation and sustainability: A review. *KSCE Journal of Civil Engineering, 20*(3), 1076–1083. https://doi.org/10.1007/s12205-016-0705-0

Moavenzadeh, F., Hanaki, K., & Baccini, P. (2002a). *Future cities: Dyanmics and Sustainabilit*. Kluwer Academic Publishers.

Moavenzadeh, F., Hanaki, K., & Baccini, P. (2002b). *Sustainable transportation in an age of global warming*. Springer.

Newman, P., & Kenworthy, J. (1999). *Sustainability and cities: Overcoming automobile dependence*. Island Press.

Organisation for Economic Cooperation and Development OECD. (2018). *Integrating urban public transport systems and cycling*. Summary and Conclusions of the International Transport Forum (ITF) roundtable on integrated and sustainable urban transport held in Tokyo, between 24–25 April, 2017.

Palmgren, E., & Zapico, J. L. (2008). *Multimedia applications for A sustainable urban lifestyle*. Ericsson.

Rivas, M. E., Suárez-Alemán, A., & Serebrisky, T. (2019). *Urban transport policies in Latin America and the Caribbean: Where we are, how we got here, and what lies ahead*. Inter-American Development Bank. United Nations (2021). UNECE NEXUS Sustainable Mobility and Smart Connectivity.

Salim, I. (2003). Security Aspects of Multimodal Transport. In *Proceedings of the Expert Meeting on the Development of multimodal Transport and Logistics services*, 24–26.

Schiller, P. L., Bruun, E. C., & Kenworthy, J. R. (2010). *An Introduction to Sustainable Transportation policy, planning, and implementation*. Earthscan, Washington DC.

SLOCAT. (2021). *Transport and climate change global status report* (2nd ed.). Section 3.3: Walking and cycling. https://tcc-gsr.com/responses-to-policies/walking-and-cycling

Smith, G., Sochor, J., & Karlsson, I. C. M, (2018). Mobility as a Service: Development scenarios and implications for public transport, *Research in Transportation Economics*, Elsevier, 69(C), 592–599.

Smith, M., Ikeda, E., Hawley, G., Mavoa, S., Hosking, J., Egli, V., Zhao, J., Mackay, L., Donnellan, N., Amann, R., Mackie, H., & Witten, K. (2020). An integrated conceptual model of environmental needs for New Zealand children's active travel to school. *Journal of Transport & Health, 16*, 2214–1405, https://doi.org/10.1016/j.jth.2019.100814

Theis, T. (2012). What is sustainability? In T. Theis & J. Tompkins (Eds.), *Sustainability: A comprehensive foundation*. Connexions. Retrieved October 1, 2012, from http://cnx.org/content/m41188/1.7/

Tønnesen, A., Knapskog, N., Uteng, T. P., & Oksenholt, K. V. (2020). The integration of active travel and public transport in Norwegian policy packages: A study on 'access, egress and transfer' and their positioning in two multilevel contractual agreements. *Research in Transportation Business & Management*. https://doi.org/10.1016/j.rtbm.2020.100546

United Nations. (2021). *Sustainable transport, sustainable development*. Interagency report for second Global Sustainable Transport Conference.

Vijayakumar, N. & Mehendiratta, G. (2011). Role of ICT in Sustainable Transportation – Focus on Reducing Traffic Congestion. An unpublished Master's Thesis in Informatics from the University of Boras. Retrieved on 25 December 2019 from https://www.diva-portal.org/smash/get/diva2:1312377/FULLTEXT01.pdf

Volker, J., & Handy, S. (2021). Economic impacts on local businesses of investments in bicycle and pedestrian infrastructure: A review of the evidence. *Transport Reviews, 41*(4), 401–431. https://doi.org/10.1080/01441647.2021.1912849

Weisers, M. (2008). *Mark Weisers homepage*. Retrieved from http://www.ubiq.com/hypertext/weiser/UbiHome

Wegener, M., Schwarze, B., Spiekermann, K., Brosch, K., Huber, F., Müller, M., & Reutter, O. (2019). Modelling the Great Transformation in the Ruhr Area. *Transportation Research Procedia, 41*, 231–239, ISSN 2352–1465, https://doi.org/10.1016/j.trpro.2019.09.042

Witting, R., & Godefrooij, T. (Eds.). (2009). *Cycling-inclusive policy development: A handbook*.

World Bank. (2021). *Investing for momentum in active mobility*. Discussion Paper on transport decarbonization investment.

World Commission on Environment and Development (WCED). (1987). *Our common future*. Oxford University Press.

World Health Organization (WHO). (2020). *Physical activity*. https://www.who.int/news-room/fact-sheets/detail/physical-activity

Zheng, W., Nie, T., Zhao, X., Li, Q., & Han, X. (2021). The configuration of transfer facilities in integrated passenger transport hubs. In CICTP 2021 (pp. 2007–2014).

Chapter 5
Smart Cities and Their Impact on Urban Transportation Systems and Development

Oluwayemi-Oniya Aderibigbe and Trynos Gumbo

5.1 Introduction

In today's rapidly growing urban areas, transportation issues have become increasingly important worldwide. The efficient functioning of transportation systems is crucial for city sustainability and livability (Ajala, 2018). However, urbanization has led to various challenges, such as increased strain on infrastructure, loss of productivity due to traffic congestion, and health problems caused by vehicle emissions (Ajala, 2018). To address these issues, the concept of the Smart City has emerged. A Smart City integrates information and communication technologies (ICTs) to create a complex system that connects citizens, businesses, transportation, communication networks, services, and utilities (Kummitha & Crutzen, 2017; Rivera et al., 2021). Smart Cities leverage digital technology and innovative approaches to enhance city services and deliver efficient systems (Rivera et al., 2021). They are characterized by the ability to sense and respond to challenges using technology and embedded information systems (Ramaprasad et al., 2017). The development of Smart Cities offers economic opportunities, enhances public safety, and improves various urban operations (Musa, 2016a, b). These cities rely on technologies such as information and communication technology (ICT), the Internet of Things (IoT) network, and geographical information systems (GIS) (Caragliu et al., 2011; Albino et al., 2015). Elements of a smart city include infrastructure, social inclusion, environmental sustainability, and business-led urban development (Caragliu et al., 2011; Albino et al., 2015).

To build a smart city, certain characteristics must be considered, including smart economics, smart mobility, smart environment, smart governance, smart people, and smart living (Gittinger et al., 2007). The concept of a smart city involves both public and private involvement to achieve its objectives (Caragliu et al., 2011). The availability of technical solutions has made it possible for each city to become smarter (Kashif et al., 2018). The design of a smart city is based on four core

O.-O. Aderibigbe et al., *Emerging Technologies for Smart Cities*,
https://doi.org/10.1007/978-3-031-66943-9_5

Fig. 5.1 An overview of smart city with smart transportation network and architecture. (Source: Mishra (2020))

concepts. First, connectivity serves as the foundational layer, enabling the collection of real-time data through smart sensors and cloud-based storage for real-time decision-making (see Fig. 5.1). Second, mobility focuses on effectively and efficiently moving people, goods, and information. The third concept is security, encompassing the improvement of public and private security, data protection, and cyber-security using the latest ICT technologies. Finally, sustainability plays a vital role in smart cities, addressing sectors such as transportation, energy consumption, climate change, utilities, security observations, and financial services.

5.2 Brief Overview of Smart City

Goncalves et al. (2021) suggest that the concept of a 'smart city'lacks a precise definition, but it generally involves the integration of technology within urban settings. Essentially, a smart city can be characterized as a technologically advanced environment that utilizes innovative technologies to connect people, information, and various aspects of the city. The aim is to create a sustainable and environmentally friendly city, promote innovative business practices, and enhance the overall quality of life for its residents (Bakici et al., 2013). Kauf (2020) also defines a smart city as a tripod, consisting of infrastructure, operations, and people, which rely on strong interconnections between each pillar to become smarter for the benefit of all. This definition highlights three fundamental characteristics inherent in a smart city: (1) instrumented, with the presence of ICT solutions (sensors, mobile devices); (2) connected, with the availability of connections between the physical and virtual worlds using instrumentation; and (3) intelligent, implying the ability to utilize new technologies in a city's development process.

Smart cities are characterized as urban environments that prioritize investments in human and social capital, alongside the development of both traditional (transport) and modern (ICT) communication infrastructure. The goal is to stimulate sustainable economic growth and enhance the overall quality of life. According to Komninos (2002), smart cities are distinguished by their capacity for learning and innovation, fostered by the creativity of the population, the presence of knowledge institutions, and the availability of digital infrastructure for effective communication and knowledge management. In a contemporary smart city, the focus is on actively listening to the needs of residents and seamlessly adapting to them. Technology plays a crucial role in providing comfort and safety, ultimately contributing to an improved quality of life. However, it is crucial to recognize that technological solutions alone cannot fully fulfill their intended purpose without considering the social context. Therefore, a comprehensive approach is necessary, involving both top-down implementation by authorities and bottom-up involvement from citizens, in order to effectively realize the vision of future cities (Robinson, 2015). In the context of the fourth industrial revolution, a smart city is be defined as the integration of information technology with infrastructure, architecture (see Fig. 5.1), and everyday objects to address social, economic, and environmental challenges (Townsend, 2013). Essentially, it entails the fusion of information and communication technology (ICT) with conventional urban infrastructure, enabling seamless coordination and integration through cutting-edge digital technologies (Batty et al., 2012).

Mitchell (2007) defines smart cities as intelligent systems formed through the effective combination of digital telecommunication networks (the nerves), embedded intelligence (the brains), sensors and tags (the sensory organs), and software (the knowledge and cognitive competence). These systems do not exist in isolation or connect solely through human intermediaries. Rather, there is a growing network of interconnected systems, from building installations to household appliances, transport systems, and utility networks, ensuring the safety and well-being of city residents. The various definitions presented highlight the lack of universally accepted concept of a smart city, as it continues to evolve due to the ongoing digitization and the fourth industrial revolution's impact on modern urbanization. From the information above, it can be inferred that technology is the driving force behind smart cities.

In summary, addressing transportation challenges and achieving sustainable urban development are critical for the efficiency and growth of cities. The concept of the Smart City, with its integration of ICTs and innovative approaches, offers solutions to enhance urban systems and improve quality of life. By leveraging technology and adopting smart policies, cities can effectively manage their resources, improve transportation networks, and create sustainable and livable environments.

As seen in Fig. 5.2, planning for a smart city involves several processes which can help in the full realization of its goals. The first and most important is the very pivotal identification by stakeholders of the problem(s) which need to be solved and

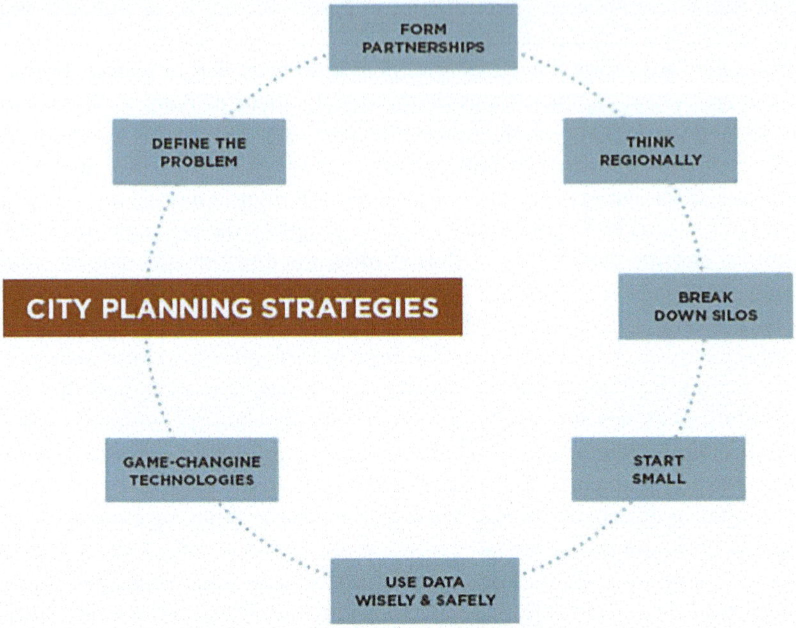

Fig. 5.2 The processes of initiating a smart city. (Source: Adapted from Synder Associates (2015))

also to form partnerships with the people. There is a need for citizen participation who are the end-users for the full achievement of a smart city. There is also a need to think regionally and internationally, thus stakeholders should not only be involved in analytical thinking of approaches and methods which have been adopted at both regional and international levels in achieving smart cities but should also apply them at local level.

Figure 5.3 provides the basic underpinnings upon which smart cities are formed. In summary, Kashif et al. (2018) summarised some of the benefits of turning our cities to smart one and this includes but not limited to.

- Smart cities leverage smart technologies, including Google Maps and similar tools, to offer users up-to-date and dynamic information on city routes and connections with neighboring cities. This enables users to access real-time information about routes, nodes, and other relevant aspects of the city's infrastructure.
- Smart cities offer road users real-time updates on traffic congestion through various formats, including text and audio/video presentations.
- Smart cities have a proven track record of assisting travelers by offering valuable information about important locations and nearby amenities. This includes details on fuel stations, restaurants, hospitals, emergency services, and much more.
- Smart cities utilize real-time data streaming analytics to offer comprehensive insights into the lives of their citizens.

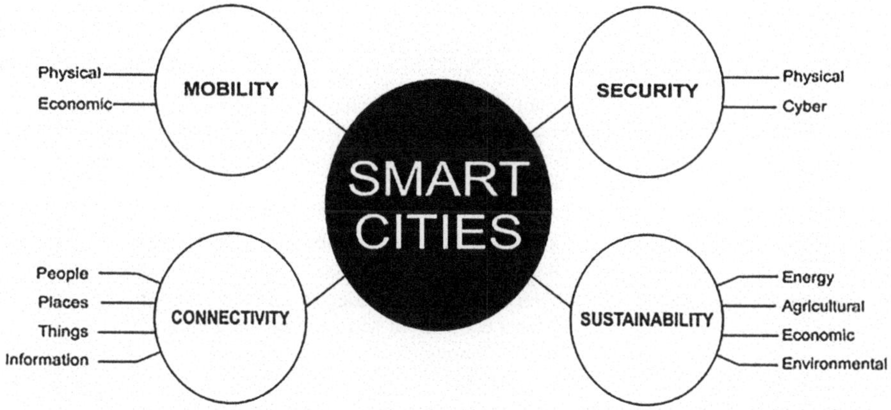

Fig. 5.3 Roadmap four pillars of a smart city. (Source: Adapted from Kashif et al. (2018))

- Users have the capability to transmit and receive emergency messages to users in different cities using technologies such as GSM/GPS and other internet services. Additionally, they can notify nearby emergency organizations through means such as emails or telephone calls.
- In smart cities, users have the convenience of easily providing or receiving emergency pickup services, healthcare assistance, information about nearby buildings, and sharing travel experiences with other travelers, particularly on highways. This is made possible through the interconnectedness of smart cities, where data tracking systems enable seamless communication and coordination.
- The reporting of any crime, security surveillance, weather forecast, or witnessed misconduct to nearby authorities is encouraged while ensuring proper privacy measures are in place.

Table 5.1 summarizes smart cities adoption with reference to major developments and approaches promoting smart development.

Table 5.1 showcases the various approaches implemented in the development of smart cities, spanning areas such as smart transportation, safety measures, and community engagement (Smith et al., 2018a, b). In the realm of transportation, these approaches encompass the utilization of electric vehicles, the promotion of active travel modes, and the integration of advanced technologies (Jones & Brown, 2019). Furthermore, Musa (2017a, b) highlights a direct and indirect positive correlation between economic development and smart city initiatives. Implementing smart projects, including intelligent traffic lights, artificial intelligence-based crime detection tools, and gun detection solutions, has contributed to a decline in crime rates within major cities of developing nations (Gupta & Sharma, 2020). These smart solutions have also led to increased employment opportunities, the emergence of start-up businesses, and enhanced citizen engagement, subsequently improving the overall quality of life (Choudhury & Abraham, 2016). The relationship between smart cities and economic development is visually depicted in Fig. 5.4.

Table 5.1 Smart city initiative and thematic areas

Thematic areas	Smart initiatives/innovations
Transportation	1. Autonomous electric shuttles 2. Smart traffic signals to address traffic congestion 3. Smart parking spaces 4. Walking and bicycling initiatives (active travel/non-motorized transport) 5. LED street light conversion projects
Safety	1. Crime mapping smart solutions to reduce crimes 2. GIS and gun detection tools
Net zero emission/green initiatives	1. Electric vehicles to reduce the carbon footprint 2. Highest densities of LEED certified buildings 3. Smart technology to achieve zero waste
Citizen involvement	1. Public Wi-Fi 2. 311 centralized services, online government services 3. Open data and open government smart initiatives

Source: Adapted from Musa (2017a, b)

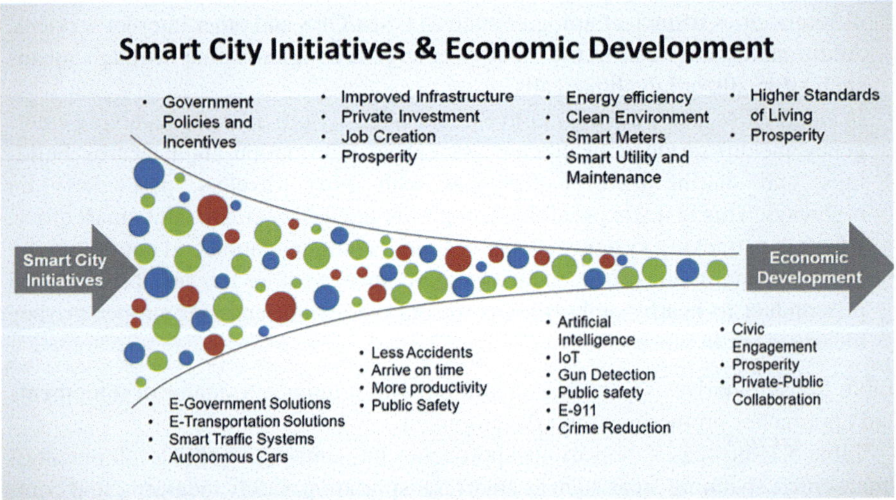

Fig. 5.4 Relationship between smart cities initiatives and economic development. (Source: Musa (2017a, b))

The pursuit of a smart city is not an end in itself, but rather a means to enhance the well-being of its citizens through intelligent approaches. Understanding how this tool can enhance quality of life is crucial. To evaluate a city's progress towards becoming a truly smart city, the Smart City Maturity Model (see Fig. 5.5) encompasses five dimensions, each representing a key aspect of the city's transformation (Yigitcanlar et al., 2018). The first four dimensions enable the assessment of capabilities required for a smart city, while the final dimension focuses on integrating smartness into various aspects of human life (Caragliu et al., 2011). This final

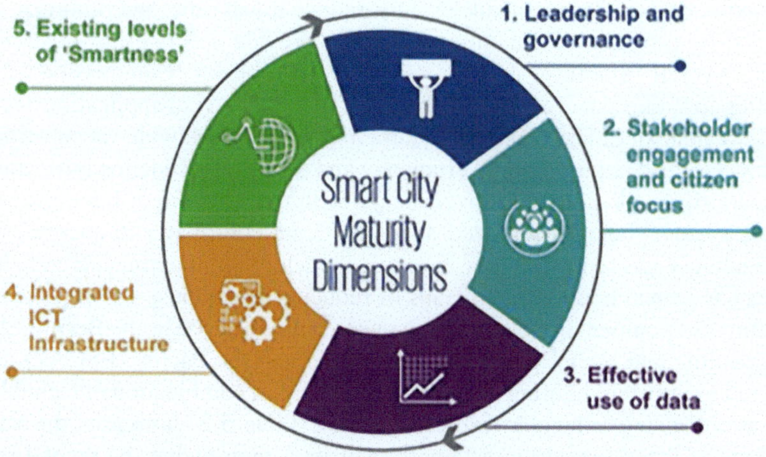

Fig. 5.5 Smart cities and digital governance. (Source: Mishra (2020))

dimension explores the interplay between ICT infrastructure, data utilization, citizen engagement, participation, and governance in driving smart initiatives within the city. It examines how these elements interact to achieve a cohesive and comprehensive smart city framework (Anthopoulos, 2017). By assessing the integration level of smartness across multiple domains, cities can gauge their progress and identify areas that require further development (Albino et al., 2015).

5.3 Intelligent Transportation Systems for Smart Development

Cities worldwide are confronted with persistent transportation issues, prompting them to invest in Intelligent Transportation Systems (ITS) as a means of mitigating these problems (Smith et al., 2018a, b). ITS leverages Information and Communications Technology (ICT) to optimize the utilization of existing transport infrastructure, aiming to enhance transport services, alleviate congestion, reduce accidents, and minimize air pollution (Chen & Ni, 2017). The emergence of smart cities, which has gained significant traction in the past two decades, has further propelled cities to adopt a comprehensive approach in leveraging smart technology to enhance urban services (Caragliu et al., 2011).

Intelligent Transport Systems (ITS) refer to the application of ICT in the transportation domain, encompassing infrastructure, vehicles, users, traffic management, mobility management, and integration with other modes of transport (Goodchild, 2007). ITS services and products are built upon the integration of diverse information about transport networks, travelers, and vehicles. These solutions can be deployed across infrastructure to optimize network management and increase productivity, on vehicles to enhance safety, reduce travel time, and minimize emissions, and by travelers to streamline ticketing, provide accurate information, and enhance overall mobility and accessibility (European Commission, 2019). According to Lombardi et al. (2012a, b), ITS represents a vital application and integral component of smart cities, and numerous cities worldwide are investing in smart transport initiatives. Energy efficiency and emissions reduction have emerged as critical drivers for smart city investments, with the adoption of intelligent transportation systems being recognized as a significant contributor to energy savings and greenhouse gas reduction. For instance, the adoption of automatic and electric vehicles has gained traction as a means of reducing greenhouse gas emissions and promoting environmental well-being (Newman-Askins, 2003a, b; Bertini, 2005). The Economic and Social Commission for Asia and the Pacific (ESCAP) (2016) conducted a study highlighting that ITS encompasses a combination of technologies aimed at enhancing vehicular traffic efficiency. Table 5.2 showcases the varying definitions of ITS adopted by different countries, emphasizing the pivotal role of technology in the adoption of intelligent transport systems for sustainable urban planning.

Table 5.2 Definitions of ITS by different countries

Countries	Definition
European Union	Applications of information and communication technology to transport
ITS United Kingdom	A combination of information and telecommunications technology that enables provision of online information in all areas of public and private administration
ITS Japan	A system that capitalizes on leading-edge information technology to support the comfortable and efficient transportation of people and goods, anticipating a "quantum leap" in safety, efficiency and comfort
ITS Canada	Applications of advanced and emerging technologies (computers, sensors, controls, communications and electronic devices) in transportation systems to save lives, time, money, energy and the environment
Republic of Korea	A transportation system that improves efficiency and safety through automated systems management and provides transportation data through services that integrate the state-of-the-art technologies as electric control and communication with vehicles and other transport infrastructure
Malaysia	Applications of advanced and emerging technologies (computers, sensors, controls, communications and electronic devices) in transportation systems to save lives, time, money, energy and the environment; and the integration of information and communication technology with transportation infrastructure, vehicles and users
Philippines	Applications of computing, electronics, information technology and communications to solve problems in all transport sectors
China	A new generation of transportation systems for improving safety, efficiency, accessibility and sustainability in transportation through application of advanced information technology

Source: Intelligent Transport System for sustainable mobility, UNECE, From the definition of Intelligent Transport Systems (ITS)

5.4 Intelligent Transportation System and Its Impact on Sustainable Development

Intelligent Transportation Systems (ITS) have proven to be highly advantageous in achieving and sustaining a well-functioning transport system by incorporating diverse Information and Communication Technology (ICT) solutions, thereby contributing to sustainable development (Agyemang, 2020). Several notable impacts of ITS on sustainable development include:

5.4.1 *Traffic Management*

Cities worldwide are making significant investments in technology to alleviate traffic congestion, leading to the implementation of various solutions (Economic and Social Commission for Asia and the Pacific, ESCAP (2016). Intelligent Transportation Systems (ITS) deployment plays a crucial role in maximizing traffic management efficiency (ESCAP, 2016). One of the notable benefits of ITS on transportation services is the reduction of traffic congestion and subsequent alleviation of delays. For instance, in the Republic of Korea, ITS services have contributed to a 15–20% increase in average traffic speed (ESCAP, 2016). Furthermore, the traffic light management system developed by the city of Irvine in California has effectively monitored traffic flow and reduced congestion (ESCAP, 2016). As depicted in Fig. 5.6, this system accurately predicts the arrival time of a group of vehicles to ensure that the upcoming traffic light is green upon their arrival. In certain cities, sensors are interconnected with all traffic lights, enabling the control of the number of halted vehicles at each junction and synchronizing the traffic light cycles to optimize traffic flow. This approach empowers cities to significantly enhance the movement of vehicles (ESCAP, 2016).

5.4.2 *Real Time Information*

The deployment of Intelligent Transportation Systems (ITS) has brought notable benefits to drivers and road users, including access to real-time traffic information regarding delays and traffic conditions (Habtemichael et al., 2018a, b). This information empowers drivers to make informed decisions and strategically avoid traffic congestion by adjusting their speed accordingly (Habtemichael et al., 2018a, b). As a result, congestion is reduced, and traffic efficiency is improved, leading to increased punctuality and reliability of public transportation services such as rail, bus, and taxis (Habtemichael et al., 2018a, b). Consequently, the public's trust in and utilization of these services grows (Habtemichael et al., 2018a, b). The enhanced traffic efficiency facilitated by ITS also enables transportation managers to respond

Fig. 5.6 Smart transportation system. (Source: Mishra (2020))

promptly to traffic incidents (Habtemichael et al., 2018a, b). By alerting drivers about potential hazards and emergencies, traffic incidents can be mitigated, further reducing congestion and preventing accidents (Habtemichael et al., 2018a, b).

5.4.3 Carbon Emission

Transportation plays a significant role in contributing to CO2 emissions and environmental burden, particularly when considering the extraction of fossil fuels as a contributor to pollution in this sector (Habtemichael et al., 2018a, b). According to data from the International Energy Agency (IEA) in 2012, transportation accounts for 22% of energy consumption related to CO2 emissions, with 74% of total CO_2 emissions in the sector stemming from road traffic (IEA, 2012). This highlights the substantial impact of exhaust emissions from vehicles on pollution and the environment, as well as the health of individuals.

To address this issue, reducing emissions from vehicles that rely on fossil fuels has become a primary objective in pollution reduction efforts. The level of particulate emissions from motor vehicle exhausts is a significant contributor to air pollution, making it a key target for mitigation strategies (Clean Air Initiative for Asian Cities Center Inc, 2012). According to ESCAP (2016), the advancement of Intelligent Transportation Systems (ITS) holds promise in reducing fuel consumption and greenhouse gas emissions through the mitigation of traffic congestion and prevention of motor vehicle idling. The Department of Transport in the Republic of Korea has adopted an ITS development strategy with the goal of achieving a 12% decrease in carbon emissions from transportation by 2020. The implementation of ITS across 1000 km of the national highway network is expected to lead to an annual reduction of 19,000 tonnes of CO_2. Furthermore, the introduction of the "High Pass" electronic toll collection service is projected to contribute to an annual reduction of 23,000 tonnes, while integrated ITS services supporting individual vehicles have the potential to yield a 15% decrease in carbon emissions (ESCAP, 2016). Furthermore, the adoption of ITS technologies, such as the Advanced Public Transportation System (APTS), enhances the punctuality and usability of public transport, leading to increased usage compared to personally owned vehicles (ESCAP, 2016). Overall, the deployment of ITS and the shift towards sustainable transportation systems play a crucial role in reducing CO_2 emissions, addressing air pollution, and promoting the use of public transportation (Habtemichael et al., 2018a, b; ESCAP, 2016).

5.4.4 *Smarter Energy*

The growing concerns regarding climate change, increasing energy prices, and the emergence of advanced technologies have prompted individuals to become more proactive in their energy consumption and make smarter choices in adopting technology (Jones et al., 2019). The integration of new smart grid technologies (see Fig. 5.7) is transforming the relationship between utilities and end users, equipping both parties with innovative tools to reduce energy consumption, costs, and greenhouse gas emissions (Smith et al., 2020). Through the implementation of smart grids, utilities can achieve several benefits:

Fig. 5.7 A smart energy system. (Source: Mishra (2020))

(1) Empower consumers: By providing timely data on energy consumption, utilities enable consumers to have better control over their energy usage. This empowers individuals to make informed decisions and effectively manage their energy consumption (Wu et al., 2018).
(2) Deliver cleaner energy: Smart grids facilitate the incorporation of renewable energy sources and electric vehicles into the power grid, leading to a reduction in greenhouse gas emissions. By optimizing the integration of clean energy sources, utilities contribute to a greener and more sustainable energy landscape (Smith et al., 2020).
(3) Improve grid reliability: Through the use of dynamic, automated, and secure networks, smart grids enhance the reliability of the power grid. Utilities can effectively manage generation assets and the distribution grid, resulting in improved operational efficiency and a more robust electricity supply system (Wu et al., 2018).

In summary, the adoption of smart grid technologies offers utilities and consumers valuable tools to reduce energy consumption, lower costs, and mitigate greenhouse gas emissions. By empowering consumers, incorporating cleaner energy sources, and improving grid reliability, smart grids contribute to a more sustainable and efficient energy future (Jones et al., 2019; Smith et al., 2020).

5.5 Intelligent Transportation System Technologies and Benefits

Intelligent transportation systems (ITS) encompass the utilization of sensing, analysis, control, and communication technologies within ground transportation to enhance safety, mobility, and efficiency (Rakha et al., 2021). By processing and exchanging information, ITS offers a diverse array of applications aimed at alleviating congestion, enhancing traffic management, reducing environmental consequences, and maximizing transportation advantages for both commercial users and the general public (Wang et al., 2020). At the heart of ITS lies the integration of information and control technologies into transportation system operations, constituting its technical foundation. This integration encompasses communication systems, automatic control mechanisms, and computer hardware and software, all working harmoniously to optimize the functioning of transportation networks (Ghaffarian Hoseini et al., 2018).

These includes:

5.5.1 On-Board Vehicle Safety Systems Description

There are various on-board intelligent transportation system (ITS) technologies that are installed in vehicles, including cars, trucks, and buses, to provide drivers with essential safety measures and precautions while driving. Deakin (2003a, b) highlighted several on-board vehicle safety measures as follows:

(1) Collision warning systems: These systems are designed to detect objects in the front or rear path of the vehicle. They alert the driver about oncoming vehicles or obstacles that may not be visible to them through the use of flashing lights and warning signals.
(2) Lane warning systems (lateral collision warning systems): This ITS technology alerts the driver when an object or vehicle is approaching from the side, particularly during lane changing or overtaking maneuvers.
(3) Minimum Headway Advisory systems: These systems inform the driver if they are following another vehicle too closely, helping to prevent accidents by maintaining a safe distance.
(4) Lane sensors: Installed on vehicles, lane sensors read pavement markings to determine the vehicle's position and heading within its lane, assisting drivers in staying within the designated lane.
(5) Night vision systems detector: Equipped with detectors, night vision systems enhance visibility in low-light conditions, helping drivers stay in their lane and detect objects that may be difficult to see.
(6) Adaptive cruise control: Adaptive cruise control allows drivers to automatically adjust their vehicle's speed to maintain a safe distance from the vehicle directly ahead, enhancing safety during highway driving.
(7) Skid control: Skid control systems assist vehicles in regaining traction and control when skidding occurs, reducing the risk of accidents caused by loss of control.
(8) Rollover control systems: These systems automatically apply brakes to help the vehicle regain stability when the possibility of tipping or rolling over is detected.

On-board safety systems have proven to be highly effective and efficient in reducing road accidents and improving overall driving performance (Deakin, 2003a, b).

5.5.2 *Changeable Message Signs or Variable Message Signs (VMS)*

Dynamic message signs (DMS) are traffic control devices that are programmed to display messages using letter symbols or a combination of symbols (Ma et al., 2017). These signs offer numerous advantages, such as delivering real-time information on changing conditions and weather updates, which in turn enhance the overall performance and efficiency of transportation systems (Yang et al., 2018). By providing timely and relevant information, DMS contribute to reducing crashes and offering travelers up-to-date details about traffic situations (Zhang et al., 2019). The range of information delivered through these intelligent transportation system (ITS) tools includes route diversions, lane change or diversions, travel speeds, traffic delays or warnings, as well as weather updates, among other valuable traffic-related data (Chang et al., 2019).

To facilitate the effective functioning of DMS, four key components play crucial roles: display boards, control centers, monitoring equipment, and communication networks (Sun et al., 2016). Display boards are responsible for presenting the messages to the drivers, while control centers oversee the management and coordination of the information displayed on the signs (Sun et al., 2016). Monitoring equipment ensures the continuous monitoring of traffic conditions, which enables the generation of accurate and timely messages (Ma et al., 2017). Lastly, communication networks establish the vital connections between the various components, allowing for the seamless flow of information within the ITS system (Yang et al., 2018).

Benefits
Variable Message Signs (VMSs) serve as effective tools for providing valuable information to road users regarding congestion, route allocation, assignment, and speed restrictions (Rahman et al., 2019). Through the delivery of advice and warnings, VMSs play a crucial role in improving routes allocation, reducing vehicle miles traveled and hours spent on the road, mitigating traffic congestion, minimizing travel times and delays, as well as preventing air and environmental pollution (Huang et al., 2018; Li et al., 2017). By informing travelers about accidents and congestion, VMSs enable them to make informed decisions, such as postponing or canceling trips, thereby reducing the number of vehicles on the road and preventing air and noise pollution (Huang et al., 2018; Rahman et al., 2019).

5.5.3 Highway Advisory Radio (HAR)

The Highway Advisory Radio (HAR) is an additional Intelligent Transportation System (ITS) tool that plays a vital role in providing real-time traffic information to travelers (Elbery et al., 2017). The HAR system functions by alerting motorists as they enter a designated HAR zone, which can be either permanent and situated in high congestion areas or temporary and employed during construction zones, traffic incidents, or special high-traffic events (Golias et al., 2020; Su et al., 2019). Its primary purpose is to deliver accurate and timely information to travelers, enabling them to modify their travel plans accordingly (Elbery et al., 2017). For example, the HAR system can promptly notify motorists about traffic conditions, such as accidents and adverse weather, allowing them to consider alternative routes (Golias et al., 2020; Su et al., 2019).

5.5.4 Pedestrian Detection Systems

Pedestrians contribute significantly to motor vehicle-related fatalities and road injuries, as highlighted by Hass-Klau (1993a). To mitigate these risks, modern transport technologies offer effective solutions to enhance pedestrian safety, particularly at signalized intersections and crossings. Among these technologies are auditory signals and symbol-based warning signs, which cater to individuals who may have difficulty reading crossing light displays (Hass-Klau, 1993b). Auditory signals are designed to inform pedestrians, especially those who cannot read the display, about the appropriate time to cross the road, typically indicated by a change in the display light. To aid pedestrians in decision-making, a countdown display is often employed, visually indicating the remaining seconds before the signal changes (Hopkinson, 1989a, b). This visual and auditory combination reinforces the safety message and assists individuals with disabilities (Hass-Klau, 1993b). Furthermore, Deakin (2003a, b) suggests that the effectiveness of Pedestrian detection systems can be improved by incorporating additional technologies such as crossing displays and motorist warnings, which can be utilized at both signalized and unsignalized intersections.

5.5.5 Signal Detectors and Signal Timing for Bicycles

Traffic signal design and operation typically focus on accommodating motorized vehicles, but it is important to consider non-motorized transport as well, such as bicycles. Detector technology used in traffic signals can also be applied to recognize the presence of bicycles, allowing for signal changes and timing adjustments that enhance safety and convenience for cyclists (Deakin, 2003a, b). This approach enables priority treatment for cyclists, granting them advantages when making left turns or navigating complex intersections. For instance, the implementation of a bike stop line, positioned ahead of the motor vehicle stop line, coupled with detectors triggering bike signals, allows cyclists to gain a head start (Deakin, 2003a, b). In areas where cycling is a prominent mode of transportation, signal timings can be coordinated to ensure that bicycles can keep pace with the overall traffic flow. DeRobertis and Wachtel (1996a, b) highlight the use of bike-sensitive detectors and signal timings to provide priority treatment for cyclists, particularly in regions with high bike usage. This approach supports and encourages non-motorized travel, fostering safer and more comfortable conditions. By promoting active travel and non-motorized transport, cities can embrace sustainable transportation practices and transition towards smart cities. Moreover, the inclusion of bike-sensitive detectors makes intersections with traffic-actuated signals more accessible and comfortable for cyclists (Deakin, 2003a, b).

5.5.6 Electronic Toll Collection (ETC)

This is an innovative payment method used at toll booths that offers enhanced speed and convenience compared to traditional manual toll collection systems. ETC enables automatic toll collection without requiring vehicles to come to a complete stop. This technological advancement contributes to the reduction of waiting times for motorists and helps prevent traffic congestion.

The process involves the installation of an electronic tag, known as a transponder, or the attachment of account information directly onto the vehicle. At the toll gate, a receiving antenna reads the information from the transponder or the vehicle-mounted device. Afterwards, the toll fee is automatically subtracted from the pre-paid toll account linked to the pass holder (Transportation Research Board, 2002).

5.5.7 Smart Snow Ploughing

In recent years, the utilization of advanced technologies has proven instrumental in assisting snowplough operators during blizzards with limited visibility. These technologies can be categorized into vehicle-installed and road-installed systems (see Figs. 5.8 and 5.9), and they encompass:

- GPS: Global Positioning System (GPS) technology offers precise information regarding the center lane and edges of the road.
- Radar Systems: Forward, rear, and side radar systems play a crucial role in identifying objects and providing both visual and audible collision warnings to operators.
- Magnetic "Nails": These specialized markers are positioned along the pavement's edge and center, aiding snowplough operators in locating the road boundaries.

Fig 5.8 A typical example of smart snowploughing equipment

Fig. 5.9 Smart snow ploughing machine used in clearing snow off roads

- Smart Tape: An innovative solution, magnetized lane-marking tape known as "Smart Tape," is detectable by equipped vehicles, assisting in lane identification and road boundary delineation.

These technologies collectively contribute to enhancing the efficiency and safety of snowplough operations in challenging weather conditions, ensuring road boundaries and obstacles are accurately identified and navigated.

5.5.8 Transit Information Systems

Transit Information Systems play a crucial role in providing real-time information to passengers regarding routes, fares, and potential delays, enabling them to plan their journeys effectively and avoid unnecessary disruptions (Chien, 2004). These systems not only offer information on transit schedules but also provide updates on the actual locations of vehicles and estimated arrival times at stations and stops. Initially, transit information was primarily provided by individual operators, but there is now a growing trend of integrating this information into multimodal traveler information systems (Chien, 2004).

To gather advanced transit information such as vehicle locations, automated vehicle locator (AVL) systems are commonly utilized on buses and trains. These systems transmit real-time data on vehicle locations to a central processor, which subsequently disseminates the information to system managers and transit users through various channels such as website displays and telephone services (Chien, 2004). By utilizing transit information systems, challenges associated with obtaining and interpreting printed transit routes are mitigated, thereby enhancing fleet management processes.

Furthermore, schedule information has been identified as a significant barrier to widespread transit use. Transit information systems address this concern by providing real-time updates, improving convenience and making public transit a more attractive option for travelers (Cats et al., 2016). The provision of real-time information helps alleviate uncertainties about arrival times and reduces frustrations arising from unexplained delays. By enhancing transparency and reducing uncertainties, transit information systems contribute to a more satisfactory user experience.

Considering the challenges faced in transportation, including environmental issues like air pollution, the adoption and integration of Intelligent Transport System (ITS) technologies become increasingly necessary (Adams and Gerike, 2016). ITS technologies offer innovative solutions for traffic management and transport planning, supporting sustainable and efficient transportation systems.

5.6 Intelligent Transport Systems and Its Impact on Travel Behaviour and Transport Planning

The implementation of Intelligent Transportation Systems (ITS) offers significant benefits to passengers, particularly in reducing travel times and improving the efficiency of their journeys. ITS facilitates this by enabling passengers to make informed decisions regarding the most suitable means of transportation and optimal routes to their destinations (Hall, 1998). One of the key ways in which ITS contributes to reducing travel and waiting times is by effectively addressing traffic congestion. Through the utilization of advanced traffic control signals and route assignment strategies, ITS helps in managing and alleviating congestion on road networks (Hall, 1998). By optimizing signal timings and dynamically assigning routes, ITS systems enhance traffic flow and minimize delays, thereby reducing overall travel times.

Furthermore, ITS supports drivers in navigating congested areas by providing them with real-time traffic information (Gordon et al., 2007). This information enables drivers to make informed decisions about route selection, allowing them to avoid traffic bottlenecks and choose alternative routes that are less congested and more time-efficient. Additionally, ITS facilitates the seamless and automated payment of tolls through Electronic Toll Collection (ETC) systems, eliminating the need for drivers to stop and manually pay tolls, thereby further reducing travel times (Hall, 1998). In conclusion, the deployment of ITS plays a crucial role in minimizing travel and waiting times for passengers. By employing advanced traffic control measures, providing real-time traffic information, and enabling automated toll payments, ITS significantly contributes to the reduction of congestion and the optimization of travel routes, enhancing overall efficiency in transportation.

Prevention of Traffic Accidents

The implementation of Intelligent Transportation Systems (ITS) has proven effective in preventing a significant number of traffic accidents, thereby enhancing safety for pedestrians and motorists. ITS offers various features and technologies that contribute to minimizing collisions and their associated consequences.

For instance, ITS plays a crucial role in preventing accidents at intersections by employing advanced signal control and detection systems. By accurately detecting the presence of vehicles and pedestrians, ITS can effectively manage traffic flow, coordinate signal timings, and implement measures to reduce the risk of collisions (ITS Asia-Pacific, 2013a, b). This proactive approach significantly reduces the occurrence of accidents and enhances safety at intersections.

Furthermore, ITS contributes to the prevention of accidents during highway merging. Through the use of intelligent systems and real-time traffic information, ITS enables drivers to make informed decisions when merging onto highways, reducing the likelihood of collisions and improving overall safety (ITS Asia-Pacific, 2013a, b). The integration of ITS technologies in highway management systems allows for better coordination and control of traffic flow, minimizing potential conflicts and enhancing safety for all road users.

Rear-end collisions, which are a common type of accident, can also be mitigated through the deployment of ITS. Advanced driver assistance systems, such as collision warning systems and automatic emergency braking, are examples of ITS applications that help prevent rear-end collisions by providing timely alerts and assisting drivers in avoiding or mitigating potential collisions (ITS Asia-Pacific, 2013a, b). By offering these safety features, ITS significantly reduces the number of accidents, injuries, and property damages caused by rear-end collisions.

Moreover, ITS contributes to promoting the safety of active mobility users, such as cyclists, through the deployment of signal detectors specifically designed for their needs. By detecting the presence of cyclists at intersections, ITS can improve their visibility to motorists and reduce the risk of crashes, thereby encouraging the use of non-motorized transport options and enhancing overall safety (ITS Asia-Pacific, 2013a, b).

In conclusion, the implementation of ITS has proven effective in preventing a wide range of traffic accidents. Through advanced signal control, real-time traffic information, and targeted safety measures, ITS significantly reduces the occurrence of collisions, injuries, and fatalities, promoting safer road environments for all users.

Comfort

Comfort is a crucial aspect of a well-functioning transport system, and Intelligent Transportation Systems (ITS) can contribute to enhancing the comfort of commuters. By providing information about the integration of different transport modes, ITS enables passengers to plan their trips more efficiently and select the most suitable modes for their journeys. This integration of transport modes, known as multi-mode transport, facilitates smoother transfers between different modes of transportation, resulting in a more comfortable and seamless travel experience. The availability of information on optimal transfers and connections between various transport modes allows passengers to make informed decisions about their routes and modes of travel. This ensures that commuters can choose the most convenient and efficient options to reach their destinations, reducing the stress and inconvenience associated with transferring between different modes of transport.

Furthermore, ITS technologies enable the automatic payment of road-use and parking fees, eliminating the need for manual transactions. This automated payment system enhances convenience and comfort for road users by reducing the time spent on administrative tasks and facilitating smoother passage through toll booths or parking facilities. By streamlining the payment process, ITS contributes to a more comfortable and hassle-free transportation experience. In summary, ITS plays a significant role in enhancing the comfort of commuters by providing information about multi-mode transport options and facilitating seamless transfers between different modes of transportation. Additionally, the automatic payment of road-use and parking fees further contributes to the convenience and comfort of road users.

Reduction in Carbon Dioxide (CO_2) Emissions

Addressing global climate change requires a significant focus on reducing carbon dioxide (CO_2) emissions, with the transportation sector being a major contributor. To achieve this goal, Intelligent Transportation Systems (ITS) tools can be deployed and alternative modes of transportation can be promoted. These initiatives have the potential to effectively decrease CO_2 emissions and mitigate the impacts of global warming. By encouraging the adoption of non-motorized transportation options like walking and cycling, as well as integrating different modes of transport, reliance on single-occupancy vehicles can be minimized.

According to a study conducted by Aderibigbe and Gumbo (2022), the adoption of non-motorized transport and the integration of various transport modes can result in a reduction in the number of vehicles on the road. This decrease in vehicle volume directly translates to lower CO_2 emissions, as fewer vehicles lead to reduced overall emissions released into the environment. Hence, the implementation of ITS tools and the promotion of alternative transportation modes are crucial in achieving a substantial reduction in CO_2 emissions, thereby playing a vital role in mitigating the environmental impact of transportation on global climate change.

Fuel Savings

Easing traffic congestion and selecting optimal routes not only improve travel times but also contribute to reducing fuel consumption and costs. Additionally, providing information on eco-driving practices further reduces fuel consumption, resulting in significant fuel savings across society. The benefits of Intelligent Transportation Systems (ITS) in transportation planning can be summarized as follows:

(1) Implementation of transport measures and policy achievement: ITS enables the implementation of various measures such as access restrictions, behavior change among users, environmentally friendly traffic management, parking

management, and priority schemes for green modes. These measures contribute to enhancing sustainability and safety in transportation systems.

(2) City infrastructure development: ITS allows cities to build upon existing infrastructure and develop innovative mobility schemes. It provides tools and measures to address urban transport challenges effectively, promoting sustainability and cost-effectiveness. Through cooperative ITS, cities can aggregate and share open data via smart city/urban mobility data portals. Open ITS architecture and public-private contractual arrangements facilitate data exchange and utilization.

(3) Support for data-driven decision-making: ITS serves as a valuable data provider for supporting, developing, monitoring, assessing, and evaluating sustainable transport initiatives. It facilitates the collection, storage, exchange, and analysis of digital information and real-time data on traffic and people flows. This data-driven approach enhances the understanding of transportation patterns and supports evidence-based planning and implementation. By leveraging the capabilities of ITS, transportation systems can achieve improved efficiency, sustainability, and safety while promoting data-driven decision-making processes.

In conclusion, the concept of smart cities is increasingly embraced by stakeholders, governments, and the public due to their utilization of advanced technologies to optimize critical systems such as electricity, water, telecommunications, and transportation. These smart cities aim to create eco-friendly urban areas that promote energy efficiency and reduce carbon dioxide emissions. The establishment of such cities worldwide is expected to contribute to the development of low-carbon societies and address the environmental challenges that negatively impact urban environments and human well-being.

References

Adams, R., & Gerike, R. (2016). Transport and climate change: Simulating the options for carbon reduction in London. *Journal of Transport Geography, 54*, 394–402.

Aderibigbe, O.-O., & Gumbo, T. (2022). Variations in mode choice of residents prior to and during COVID-19: An empirical evidence from Johannesburg, South Africa. *Sustainability, 14*, 16959. https://doi.org/10.3390/su142416959

Agyemang, E. (2020). "Uber is here to stay": Exploring the policy implications of the Uber-Local taxis turf war in Accra, Ghana, Case Studies on Transport Policy, Volume 8, Issue 1, 2020, Pages 59–66. https://doi.org/10.1016/j.cstp.2019.12.005

Ajala, A. T. (2018). Conceptualising Smart City for the development of the Nigeria's urban transportation. *Advances in Multidisciplinary and Scientific Research, 4*(2), 65–72.

Albino, V., Berardi, U., & Dangelico, R. M. (2015). Smart cities: Definitions, dimensions, performance, and initiatives. *Journal of Urban Technology, 22*(1), 3–21.

Anthopoulos, L. (2017). Understanding the smart city domain: A literature review. In E. Tambouris, A. Macintosh, & H. J. Scholl (Eds.), *Electronic participation* (pp. 13–24). Springer.

Bakıcı, T., Almirall, E., & Wareham, J. (2013). A smart city initiative: The case of Barcelona. *Journal of the Knowledge Economics., 4*(1), 135–148.

Batty, M., Axhausen, K., Giannotti, F., Pozdnoukhov, A., Bazzani, A., Wachowicz, M., Ouzounis, G., Portugali, Y., (2012), Smart cities of the future. *The European Physical Journal Special Topics, 214*, 481–518.

Bertini, R., Monsere, C., & Yin. (2005). *Benefits of intelligent transportation systems technologies in urban areas: A literature review*. Portland Statement University Center for Transportation Studies.

Caragliu, A., Del Bo, C., & Nijkamp, P. (2011). Smart cities in Europe. *Journal of Urban Technology, 18*(2), 65–82.

Cats, O., Loutos, G., Susilo, Y. O., & Reichenbach, M. (2016). Real-time information and traveler behavior: The effects of real-time information on traveler behavior and its implications for travel demand modeling. *Transportation Research Part A: Policy and Practice, 88*, 1–13.

Chang, X., Xiao, G., Li, P., Wang, Y., Zhang, K., & Liu, J. (2019). Study on variable speed limit based on traffic flow theory and big data. *Journal of Advanced Transportation, 2019*, 1–16. https://doi.org/10.1155/2019/5067163

Chen, C. H., & Ni, D. (2017). Using an intelligent transportation system to design a car-following control. *Discrete Dynamics in Nature and Society, 2017*, 1–10.

Chien, S. I. (2004). Real-time transit information systems: Evolution, current status, and future developments. *Journal of Public Transportation, 7*(5), 1–21.

Choudhury, P., & Abraham, A. (2016). Enhancing the quality of urban life using internet of things and cloud computing. In *In 2016 International Conference on Advances in Computing, Communications and Informatics (ICACCI)* (pp. 1466–1470). IEEE.

Clean Air Initiative for Asian Cities Center Inc. (2012). *Accessing: Air Pollution and Greenhouse Gas Emissions Indicators for Road Transport and Electricity*. http://cleanairinitiative.org/portal/node/11573. Accessed Feb 2015.

Deakin, E. (2003a). *Intelligent transportation systems: A compendium of technology summaries*. University of California Transportation Center, UC Berkeley. http://escholarship.org/uc/item/8fj9168f

Deakin, E. A. (2003b). Towards a safer walking environment. *Traffic Engineering and Control, 44*(6), 198–202.

DeRobertis, M., & Wachtel, A. (1996a). *Traffic calming: Do's and don'ts to encourage bicycling* (pp. 498–503). Institute of Transportation Engineers, Compendium of Technical Papers for the 66th ITE Annual Meeting.

DeRobertis, M., & Wachtel, A. (1996b). Development of guidelines for bicycle detectors. *Transportation Research Record, 1538*, 1–9.

Economic and Social Commission for Asia and the Pacific, ESCAP. (2016). *Intelligent transportation system for sustainable development in Asia and the Pacific*. Working Paper by the Information and Communications Technology and Disaster Risk Reduction Division. http://www.unescap.org

Elbery, A., Cheng, D., & Rakha, H. (2017). An evaluation of driving behavior and fuel consumption under eco-cooperative adaptive cruise control. *Transportation Research Part C: Emerging Technologies, 80*, 101–115. https://doi.org/10.1016/j.trc.2017.04.015

European Commission. (2019). *Intelligent transport systems*. Retrieved from https://ec.europa.eu/transport/themes/its_en

Ghaffarian Hoseini, A., Zhang, M., Ghaffarian Hoseini, A., Nwaogu, U. G., & Mosavi, A. (2018). Intelligent transportation systems (ITS) for urban traffic congestion management: A systematic review. *IEEE Access, 6*, 36611–36627. https://doi.org/10.1109/ACCESS.2018.2857592

Gittinger, R., Fertner, C., Kramar, H., Kalasek, R., Pichler-Milanovic, N., & Meijers, E. (2007). *Ranking of European medium-sized cities*. Final Report, Center of Regional Science, Vienna, UT.

Golias, M. M., Stathopoulos, A., & He, S. Y. (2020). Driver compliance to advisory speed limits and dynamic message signs. *Transportation Research Part C: Emerging Technologies, 120*, 102795. https://doi.org/10.1016/j.trc.2020.102795

Goncalves, G. D. O., Filho, W. L., Neiva, S. D., Deggau, A. B., Verras, M. D. O., Ceci, F., Lima, M. A., & Guerra, J. B. S. (2021). The impact of the fourth industrial revolution on smart and sustainable cities. *Sustainability, 13*, 7165. https://doi.org/10.3390/su13137165

Goodchild, M. (2007). Citizens as Sensors: The World of Volunteered Geography. *GeoJournal 69*(4):211–221. https://doi.org/10.1007/s10708-007-9111-y

Gordon, M., Clark, S., & Nee, M. (2007). *Intelligent transport systems*. The University of Auckland. Retrieved from http://www.civil.auckland.ac.nz/Assets/Departments/Transportation/MoTIFS%202007/MoTIFS_2007_Gordon.pdf.

Gupta, R., & Sharma, S. (2020). Smart city initiatives for sustainable urban development. In *Green computing strategies for competitive advantage and business sustainability* (pp. 1–18). IGI Global.

Habtemichael, F. G., Zhang, Y., & Taeihagh, A. (2018a). Understanding the effects of smart city innovations on reducing urban energy consumption: A review. *Journal of Cleaner Production, 172*, 4028–4038. https://doi.org/10.1016/j.jclepro.2017.10.194

Habtemichael, F., Xia, J., & Haugen, K. (2018b). Intelligent transportation systems: A literature review and some considerations for the future. *Journal of Traffic and Transportation Engineering (English Edition), 5*(5), 461–482. https://doi.org/10.1016/j.jtte.2018.05.002

Hall, R. W. (1998). A vision for the next century: Traveler information systems. *Transportation Research Part C: Emerging Technologies, 6*(1–2), 7–22. https://doi.org/10.1016/S0968-090X(97)00018-7

Hass-Klau, C. (1993a). Impact of pedestrianization and traffic calming on retailing: A review of the evidence from Germany and the UK. *Transport Policy, 1*(1).

Hass-Klau, C. (1993b). Traffic calming: State of the art. In *Proceedings of seminar F* (Vol. 1, pp. 107–120). PTRC Education and Research Services Ltd.

Hopkinson, P. G. (1989a). Development of traffic calming in a heavily-trafficked shopping street. *Traffic Engineering & Control, 30*(10), 482–486.

Hopkinson, P. G. (1989b). Traffic signal installation and control for the visually handicapped pedestrian. *Traffic Engineering and Control, 30*(7), 388–390.

Huang, S., Yu, X., & Zhou, W. (2018). A coordinated dynamic pricing and adaptive routing strategy for road networks under recurrent congestion. *Transportation Research Part B: Methodological, 108*, 1–24. https://doi.org/10.1016/j.trb.2017.12.011

ITS Asia-Pacific. (2013a). *ITS Guideline for Sustainable Transport in the Asia-Pacific region*

ITS Asia-Pacific. (2013b). *ITS Benefits: Safer roads and increased road safety*. Retrieved from https://www.itsap.org/wp-content/uploads/2017/03/ITSAP_IB_ITS_Benefits.pdf

Jones, C. S., & Brown, C. G. (2019). *Smart cities: A policy handbook for local government*. Routledge.

Jones, A., Smith, B., & Johnson, C. (2019). *Smart grids: An introduction*. John Wiley & Sons.

Kashif, I., Muhammad, A. K., & Areej, F. (2018). Intelligent transportation system (ITS) for smart-cities using Mamdani fuzzy inference system. *International Journal of Advanced Computer Science and Applications, 9*(2), 94–105.

Kauf, S. (2020). Smart City in the Era of the Fourth Industrial Revolution. Scientific Papers of the Silesian University of Technology, Organisation and Management Series no. 145. https://doi.org/10.29119/1641-3466.2020.145.16

Komninos, N. (2002). *Intelligent cities: Innovation, knowledge systems and digital spaces*. Spon Press.

Kummitha, R. R., & Crutzen, N. (2017). How do we understand smart cities? An evolutionary perspective. *Cities, 67*, 43–52. https://doi.org/10.1016/j.cities.2017.04.010

Li, X., Feng, T., & Li, Y. (2017). Traffic assignment with explicit consideration of routing strategies. *Journal of Advanced Transportation, 2017*, 1–12. https://doi.org/10.1155/2017/9509725

Lombardi, P., Giordano, S., & Farouh, H. (2012a). An integrative definition of smart city: A systemic literature review. In *In 2012 6th International Conference on Theory and Practice of Electronic Governance* (pp. 28–30). IEEE.

Lombardi, P., Giordano, S., Farouh, H., & Yousef, W. (2012b). Modeling the smart city performance. *Innovation: The European Journal of Social Science Research, 25: 2*, 137–149.

Ma, Z., Bie, Y., & Jia, Z. (2017). Dynamic message sign control optimization for urban expressway with variable speed limits. *Transportation Research Part C: Emerging Technologies, 80*, 127–143. https://doi.org/10.1016/j.trc.2017.03.008

Mishra, M. K. (2020). *Smart cities in the era of globalisation, ZBW—Leibniz information centre for economics*. Kiel. http://hdl.handle.net/10419/216813

Mitchell, M. (2007). *Intelligent city*. UOC Papers. e-Journal of the Knowledge Society, Retrieved from: https://pdfs.semanticscholar.org/6c8c/d3f7e497c7ee75c6c54c737e84cec5f78418.pdf. Accessed 21 Oct 2019

Musa, S. (2016a). Smart cities—A roadmap for development. *Journal of Telecommunications System & Management, 5*(3) 1000144th ser., 1–3. https://doi.org/10.4172/2167-0919.1000144

Musa, S. (2016b). Smart city applications and requirements. *International Journal of Advanced Computer Science and Applications, 7*(6), 511–515.

Musa, S. (2017a). Smart cities: Recent trends and future direction in Malaysia. *Journal of Information Systems Research and Innovation, 9*(1), 39–46.

Musa, W. (2017b). *The impact of smart cities initiatives on cities local economy development*. Master's Theses 501. https://scholars.fhsu.edu/theses/501

Newman-Askins, R. (2003a). In V. Jaeger (Ed.), *Intelligent transport system evaluation: From theory to practice*. Proceedings 21st ARRB and 11th REAAA Conference, Cairns.

Newman-Askins, W. (2003b). Intelligent transportation systems as a policy tool for reducing greenhouse gas emissions. *Transportation Research Part D: Transport and Environment, 8*(1), 1–12.

Rahman, M. A., Ashique, A., Karim, M. M., & Sadique, M. R. (2019). Optimization of variable message sign placement on highways using a hybrid metaheuristic algorithm. *Journal of Advanced Transportation, 2019*, 1–18. https://doi.org/10.1155/2019/9258613

Rakha, H., Nakamura, H., & Elbery, A. (2021). Intelligent transportation systems. In *Handbook of transportation science* (pp. 351–388). Springer. https://doi.org/10.1007/978-3-030-54717-0_10

Ramaprasad, A., Sanchez-Ortiz, A., & Syn, T. (2017). *A unified definition of a smart city* (pp. 13–24). International Federation for Information Processing IFIP, Springer International Publishing AG.

Rivera, R., Amorim, M., & Reis, J. (2021). Public transport systems and their impact on sustainable smart cities: A systematic review. In A. M. Tavares Thomé et al. (Eds.), *Industrial engineering and operations management*. Springer Proceedings in Mathematics & Statistics 367. https://doi.org/10.1007/978-3-030-78570-3_3

Robinson, R. (2015). Reclaiming the Smart agenda for fair human outcomes enabled by technology. *The Urban Technologist*, Retrieved from: https://theurbantechnologist.com/2015/03/20/reclaiming-the-smart-agenda-for-fair-human-outcomes-enabled-by-technology, 03.10.2019.

Smith, J., Anderson, C., Hart, R., King, T., & Scott, M. (2018a). Smart cities and infrastructure: A review of the UK's current policy landscape. *Energy Policy, 123*, 575–582.

Smith, M., Muir, A., & Hill, G. (2018b). *Intelligent transport systems: Recent studies and emerging applications*. In Handbook of Research on Intelligent Transport Systems.

Smith, J., Brown, L., & Thompson, R. (2020). *Smart grid technology and applications*. Academic.

Su, J., Li, M., Xu, P., Liu, T., & Li, H. (2019). A dynamic stochastic user equilibrium assignment model considering the congestion propagation effect of traffic incidents. *Transportation Research Part C: Emerging Technologies, 107*, 164–183. https://doi.org/10.1016/j.trc.2019.07.004

Sun, J., Chen, B., & Zhang, X. (2016). Optimal control of dynamic message signs for congestion management on urban expressways. *Transportation Research Part C: Emerging Technologies, 72*, 59–76. https://doi.org/10.1016/j.trc.2016.09.012

Synder Associates. (2015). *Planning for the transportation system of a smart city*. https://www.snyder-associates.com/planning-for-the-transportation-system-of-a-smart-city/

International Energy Agency, IEA statistics. (2012). *CO_2 emissions from fuel combustion—highlights*. Available http://www.iea.org/publications/freepublications/publication/CO2emissionfromfuelcombustionHIGHLIGHT S.pdf

Townsend, A. (2013). *Smart cities: Big data, civic hackers, and the quest for a new utopia*. Norton & Company.

Transportation Research Board. (2002). Electronic toll collection. In *Transit capacity and quality of service manual* (3rd ed., pp. 4–24). National Academies Press.

Wang, D., Han, X., Li, W., Li, J., & Jiang, J. (2020). Intelligent transportation systems: Recent advances and prospective future. *IEEE Access, 8*, 21494–21513. https://doi.org/10.1109/ACCESS.2020.2969383

Wu, J., Liu, W., & Wetzstein, W. (2018). Economic, environmental, and social benefits of the smart grid: A literature review. *Energies, 11*(9), 2343. https://doi.org/10.3390/en11092343

Yang, L., Liu, Y., Chen, X., Zhao, Y., & Zheng, Y. (2018). Traffic congestion prediction and control algorithm based on Dijkstra's shortest path algorithm and dynamic message signs. *Journal of Advanced Transportation, 2018*, 1–9. https://doi.org/10.1155/2018/9324680

Yigitcanlar, T., Kamruzzaman, M., & Buys, L. (2018). Understanding the smart city domain: A bibliometric analysis. *Cities, 81*, 237–247.

Zhang, Y., Chen, Q., Lai, X., Wang, D., & Jiang, R. (2019). A queueing model for analyzing the impact of dynamic message sign system on urban traffic congestion. *Physica A: Statistical Mechanics and its Applications, 526*, 120815. https://doi.org/10.1016/j.physa.2019.120815

Chapter 6
Transportation Technologies and Transportation Management

Oluwayemi-Oniya Aderibigbe, Trynos Gumbo, and Samson Olawale Fadare

6.1 Introduction

Despite the emergence of various telecommunication methods aimed at addressing transportation problems, such as telecommuting and teleshopping, the impact on combating urban transportation issues has been limited. As we progress through the twenty-first century, it is expected that transportation issues will continue to be a prominent topic in discussions and decision-making processes within both the private and public sectors. With our interconnected global community and high population density, there is a growing need for transportation systems that are efficient, adaptable, and environmentally sustainable. However, the increasing number of vehicles owned by individuals has led to rising traffic congestion, especially in developing countries. One significant contributor to traffic congestion is the long waiting times at signal lights, which can also pose a threat to emergency vehicles. Currently, there is a lack of mechanisms for clearing traffic in emergency situations. Manual traffic control and predefined timings for traffic lights are inefficient in addressing congestion issues. The primary goals of traffic congestion control and management systems are to clear traffic as quickly as possible and prioritize the passage of emergency vehicles (Chandana et al., 2017a, b). Furthermore, inefficient signal timing, where the green signal remains active for the same duration regardless of the varying traffic volumes on different lanes, results in wasted time and resources (Miller & Shaw, 2001a, b).

To tackle these urban transport challenges, advanced technologies and software that are less prone to human errors need to be adopted. The utilization of Geographic Information Systems (GIS), Unmanned Aerial Vehicles (UAV), Big Data, and the Internet of Things (IoT) has contributed to enhanced efficiency, cost savings, and an improved quality of life. These technologies have undergone significant advancements, resulting in various benefits for society such as increased efficiency, cost reductions, and an improved quality of life. Many countries have already started

O.-O. Aderibigbe et al., *Emerging Technologies for Smart Cities*,
https://doi.org/10.1007/978-3-031-66943-9_6

transforming their cities into smart cities by incorporating advanced technologies like UAVs and GIS (Mohammed et al., 2014a, b). According to Khan et al. (2018a, b), the adoption of these technologies has become necessary for managing and controlling the growing transportation externalities such as traffic congestion, accidents, and inadequate transport infrastructure. The objective of this study was to explore the use of emerging technologies in traffic management and control to achieve the overarching goal of developing smart and sustainable cities with efficient transportation systems for their residents.

6.2 Concept of Sustainable Transportation and Traffic Management

This is the development which meets the needs of the present without jeopardizing the ability of future generations to meet their own needs (United Nations, 2015a, b). Sustainable transportation revolves around economic, social, and environmental factors, which can be further categorized as follows:

- Promoting and enhancing more efficient and environmentally friendly modes of transportation.
- Improving the efficiency of the transportation process.
- Introducing new technologies.

From the aforementioned points, it is evident that achieving sustainable development in any city or nation necessitates the comprehensive development and management of the transportation system. One approach to achieving this is through the implementation of new technologies like Unmanned Aerial Vehicles (UAVs) for traffic management and control. The United Nations (2015a, b) outlines key strategies for attaining sustainable transportation and enhanced traffic management:

1. Developing integrated transportation systems that prioritize efficiency, safety, and environmental considerations.
2. Encouraging the use of public transportation, walking, and cycling by providing supportive infrastructure and amenities.
3. Promoting the adoption of low-emission and fuel-efficient vehicles, as well as alternative fuels.
4. Implementing intelligent transportation systems and technologies to enhance traffic management, such as UAVs, Geographic Information Systems (GIS), and Big Data.
5. Enhancing the efficiency of freight transportation through improved logistics and innovative solutions.
6. Encouraging land-use planning that promotes compact and mixed-use developments, reducing the need for long-distance travel.
7. Adopting pricing mechanisms and policies that internalize the social and environmental costs of transportation.
8. Fostering public awareness and education on sustainable transportation choices.

By implementing these strategies, cities and nations can work towards achieving sustainable transportation and effective traffic management, contributing to the overall goal of sustainable development.

Improved System Operation: Improved system operation, such as the application of Intelligent Transport Systems (ITS) and the integration of transport modes, is the focal point of this study, aiming to enhance traffic management and control for sustainable transportation development. The utilization of Unmanned Aerial Vehicles (UAVs) will provide real-life data on the causes of traffic congestion in the study area, specifically identifying vehicles that contribute significantly to congestion in Johannesburg.

Demand Management (Staggered Working Hours): Demand management, particularly staggered working hours, plays a vital role in mitigating traffic issues. Once real-life information on the causes of traffic problems is acquired, implementing appropriate traffic management techniques becomes necessary to reduce or eliminate the root causes. Embracing virtual work arrangements and remote shopping can significantly decrease the number of cars on the road, alleviating congestion.

Economic Instruments (Tax, Road Pricing): Economic instruments, such as taxation and road pricing, are essential in policy formulation for effective traffic management at the national level. Measures like road pricing and taxes are crucial to discourage drivers from utilizing busy routes, prompting them to opt for less congested alternatives. This approach facilitates smoother traffic flow on heavily congested roads.

Adoption of New and Improved Technologies: Adopting new and improved technologies is another pivotal method for enhancing sustainable transportation and traffic management. In recent times, there has been a shift from conventional traffic management methods to the use of UAVs for capturing real-time traffic information. UAVs offer a faster, more reliable, and less error-prone system. Considering the aforementioned discussions, it is evident that achieving a smart city requires significant improvements in transportation planning approaches.

6.3 Concept of Smart City

The concept of a "Smart City" is not fully comprehended due to the limited number of studies conducted on this topic. However, with the continuous advancements in technology, information systems, and communications, there are several key components that play a significant role in understanding and defining the concept of a smart city. According to Mohammed et al. (2014a), there are eight core factors that influence smart city initiatives.

- Management and Organization: The alignment of management and organizational goals is crucial for the effective and efficient functioning of smart cities.
- Technology: Smart cities rely on smart computing technologies that are applied to critical infrastructure components and services. These technologies integrate hardware, software, and network technologies to provide real-time awareness and advanced analytics for intelligent decision-making.
- Governance: Smart city governance involves the implementation of processes where constituents exchange information based on rules and standards to achieve common goals. Collaboration, communication, leadership, and data exchange are essential for effective smart city governance.
- Policy Context: The policy context is vital for understanding the appropriate use of information systems. It addresses institutional and non-technical urban issues and creates conditions that enable urban development.
- People and Communities: Smart city initiatives aim to engage and involve city members in governance and management, allowing them to become active participants. Their involvement can influence the success or failure of smart city initiatives.
- Economy: Economic competitiveness is a significant driver of smart city initiatives. A city with a high degree of economic competitiveness is considered to possess one of the characteristics of a smart city. Business and job creation, workforce development, and productivity improvements are some of the outcomes.
- ICT Infrastructure: The implementation of a robust ICT infrastructure is fundamental to the development of a smart city. Factors such as availability and performance are critical considerations.
- Natural Environment: Smart cities focus on increasing sustainability and enhancing natural resource management. The protection of natural resources and related infrastructure is of utmost importance.

In light of the challenges faced by conventional cities, especially in transportation due to increased car ownership and poor traffic management, it is imperative for them to transform into smart cities. This transformation can be achieved by embracing new and advanced technologies to improve efficiency, reduce costs, and enhance the quality of life.

6.4 Adoption of Unmanned Aerial Vehicles (Drones) for Traffic Management

Unmanned Aerial Vehicles (UAVs), as defined by Škorput et al. (2016), refer to aerial vehicles that operate without human control. These vehicles utilize aerodynamic forces to generate lift, can fly autonomously or be remotely piloted, and have the capability to carry either lethal or nonlethal payloads. UAVs require minimal human intervention in monitoring traffic flow within a given location or city. They offer a wide range of applications and models, which can be categorized into safety control, scientific research, and commercial applications (Mohammed et al., 2014a, b). The significance of UAVs in traffic control and management lies in their ability to determine areas or locations prone to traffic and identify vehicles contributing to congestion on major roads. By doing so, UAVs play a crucial role in reducing accidents resulting from traffic congestion. Consequently, the adoption of UAVs, also known as "drones," has gained increasing importance in traffic management and transportation planning. Azza et al. (2021a, b) emphasize that drones are extensively used for surveillance, mapping, and remote sensing. The growing interest in UAVs stems from various factors such as their low risk to human operators, ease of use, tracking capabilities, and high-resolution imaging, leading to their increasing adoption across various fields.

In civil engineering, particularly transportation engineering, UAVs have proven applicable in monitoring and controlling traffic within urban centers (Gupta et al., 2021a, b). One of the major advantages of UAVs in traffic monitoring and management is their deployability in different locations, enabling the collection of real-time information regarding infrastructure usage, including roads, bridges, and train tracks. Moreover, UAVs have been employed in combating the COVID-19 pandemic (Kumar et al., 2021a, b). Due to their mobility, UAVs utilized for traffic monitoring can gather high-resolution data, which can be analyzed in real time. The results can then be displayed, printed, or transmitted to a central server or the cloud for further analysis. The global increase in traffic volumes has led to numerous transportation challenges, including congestion, accidents, and pollution. Consequently, there is a growing need to adopt emerging technologies like drones to provide long-term solutions (Butila & Boboc, 2022a, b).

In recent years, drones or UAVs have emerged as an alternative to the conventional manual method of traffic counting, finding applications in various traffic management scenarios. They can be utilized for monitoring weather and pavement conditions, guiding emergency vehicles, responding to incidents, measuring roadway usage, monitoring parking lot utilization, forecasting origins and destinations, providing traveler information, tracking vehicle movements at intersections, and coordinating traffic signal networks (Coifman et al., 2004a, b). The integration of UAVs with other technologies offers additional benefits. For instance, Karan et al. (2014a, b) revealed that combining UAVs with other applications enables the monitoring of environmental conditions along roadsides, tracking construction projects on highways, enhancing traffic safety and management, and assessing the performance of road maintenance structures. UAVs have predominantly been employed in transportation management applications to observe, analyze, and evaluate traffic flow and safety conditions. Moreover, the use of UAVs can enhance the effectiveness of traffic operators by enriching databases, leading to improved predictions regarding congestion phenomena. UAVs facilitate quicker and more efficient interventions at challenging scenes, thereby helping to prevent accidents. Additionally, they increase the frequency of near-real-time data collection and monitoring, while their ability to cover large areas with a single drone surpasses the capabilities of single inductive or pneumatic sensors (Brahimi et al., 2020a, b). In addition to this,

UAVs play a crucial role in traffic monitoring, management, and control, offering numerous advantages over traditional methods. Their autonomy, mobility, and high resolution data collection capabilities make them valuable tools in addressing transportation challenges. By leveraging UAV technology, traffic operators can make better-informed decisions and implement effective strategies to mitigate congestion, accidents, and pollution. The real-time information collected by UAVs can be utilized to optimize traffic flow, enhance emergency response, and improve overall transportation system performance.

Furthermore, the integration of UAVs with advanced technologies and data analysis techniques opens up new possibilities for traffic management. For example, machine learning algorithms can be applied to the collected data to identify patterns, predict traffic patterns, and optimize signal timings. This enables proactive and adaptive traffic control strategies that respond to changing conditions in real time. UAVs also contribute to the development of intelligent transportation systems by providing valuable input for traffic simulation models, allowing for more accurate predictions and evaluations of various scenarios (Karla et al., 2021). However, the widespread adoption of UAVs in traffic management does come with challenges. Issues such as regulatory frameworks, privacy concerns, airspace management, and public acceptance need to be addressed to ensure safe and responsible UAV operations. Collaborative efforts between government agencies, researchers, and industry stakeholders are essential to establish guidelines, standards, and protocols for the integration of UAVs into existing traffic management systems (Kazagli et al., 2022).

In conclusion, UAVs have revolutionized the field of traffic monitoring, management, and control. Their autonomous capabilities, high-resolution imaging, and mobility enable them to collect real-time data, analyze traffic patterns, and support decision-making processes. By leveraging the advantages of UAV technology, transportation agencies can enhance the efficiency, safety, and sustainability of their networks. Continued research, technological advancements, and collaborations are crucial in unlocking the full potential of UAVs for future smart transportation systems.

6.4.1 Components of a Drone

The components of a drone can be classified into three layers: the aerial platform for flight, the ground control station for remote or autonomous operation, and the communication system for seamless coordination between the components (Giordan et al., 2020). Different authors offer varying classifications of drones based on their perspectives and frameworks. According to Tatum and Liu (2017), an unmanned aerial vehicle (UAV) or drone consists of the aircraft itself, the control system, ground and satellite-based equipment, communication links, and operators. Sanson (2019) outlines the components of a typical drone as a quadcopter, four propellers, a motor, landing gear, battery, camera, and remote control. Tatale, Anekar, Phatak, and Sarkale (2018) emphasize that the central part of a quadcopter is a lightweight and rigid frame with four arms. It incorporates a LIPO battery, four brushless DC motors (BLDC), a controller board, four propellers, a video camera, various sensors, and a light frame (see Fig. 6.1). The specific hardware components of a quadcopter may vary depending on its intended use (Ostojić et al., 2015). In general, quadcopters are equipped with a microcontroller, sensors, motors, a Global Positioning System (GPS), power supply, and telemetry devices.

6.4.2 Classification of Drones

According to Kazi et al. (2018), drones can be classified based on their control systems as autonomous or non-autonomous. Autonomous drones are controlled remotely through programmable computers using an onboard autopilot and GPS. These drones can perform tasks without human intervention. On the other hand, non-autonomous drones are manually controlled by humans using a remote controller. Structurally, drones can be categorized as Aircraft drones or Copter drones. Aircraft drones resemble planes and are equipped with wings for flying. They are primarily utilized for military and security purposes due to their high-speed capabilities. Copter drones, on the other hand, are designed like rotorcraft or helicopters, relying on spinning rotors for flight and thrust. Quadcopters are a popular type of copter drone and find applications in recreational and commercial activities, especially in densely populated urban areas (Kazi et al., 2018).

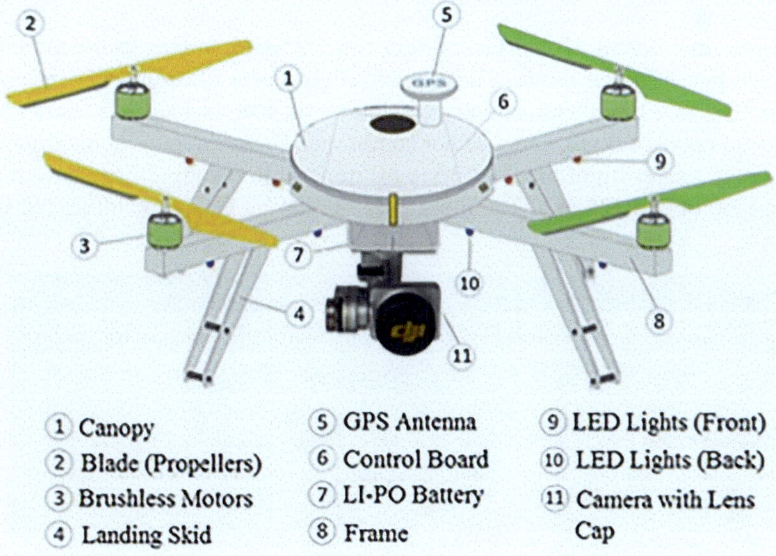

Fig. 6.1 Drone/quadcopter structure (Tatale et al., 2018)

Furthermore, drones can also be classified based on their operating mass, categorized as either heavier-than-air or lighter-than-air.

Figure 6.2 provides a vivid illustration of UAV types with a respective vehicular mass differentiation between heavier-than-air and lighter-than-air.

6.4.2.1 Classification of Drones/Unmanned Aerial Vehicles Based on Operating Mass/Weight

(i) **Heavier-than-air**: The heavier-than-air UAVs have substantial (see Fig. 6.3) vehicle mass relying on aerodynamics and propulsive thrust to fly. This classification of drones is further sub classified into *wing* or *rotor* types. Wing-type UAVs comprise fixed-wing, flying-wing, and flapping-wing methodologies to generate aerodynamic lift. On the other hand, Rotor-type UAVs include multiple rotors and propellers, which point upwards to create propulsive thrust for their lift.

(ii) **Lighter-than-air**: The lighter-than-air UAVs rely on helium gas or heated air to fly. This type of drone has an extended stamina flight capacity which can be used for photography and surveillance. One of the major challenges/disadvantages of the lighter-than-air UAVs is its inability to maneuver. Oftentimes, it requires a rope to keep it from being drifted or affected by the wind. Examples include blimps and balloons.

6.4.2.2 Classification of Drones/Unmanned Aerial Vehicles Based on Purpose of Flight/Usage

(i) Flight time versus UAV Mass: Flight time is an important factor to consider when purchasing a drone. The average flight times of drones typically range from 15 to 30 min and can vary depending on factors such as battery energy, design range of flight, and weather conditions. However, non-racing drones can achieve longer flight times of 30 to 60 min, while racing drones have a flight time of around 125 min. The flight time of a drone is largely influenced by the

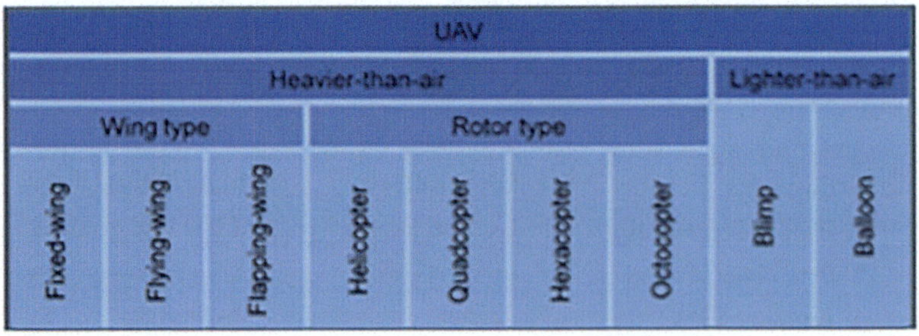

Fig. 6.2 Classification of unmanned aerial vehicles (Liew et al., 2017)

Fig. 6.3 Categories of heavier-than-air drones (Hassanalian & Abdelkefi, 2017)

capacity of its battery. Additionally, drones can vary in weight, with some being larger and heavier, while others are smaller and lighter.

(ii) Degree of Autonomy Versus Degree of Sociability: The degree of autonomy refers to the ability of drones or unmanned aerial vehicles (UAVs) to take off, perform specific tasks, and return to their base with minimal human involvement. Autonomous drones are programmed to execute tasks without the need for continuous operator intervention, representing a significant achievement in technology development. On the other hand, the degree of sociability relates to the extent to which a drone can interact or collaborate with humans. These types of drones are equipped with software that enables communication and interaction with human beings. Consequently, drones with a higher degree of sociability have a lower degree of autonomy as they require manual control for certain operations (Liew et al., 2017).

6.5 Types of Drones

Unmanned Aerial Vehicles provide an easy, effective and more efficient way of monitoring traffic. These drones have different uses and applications. The summary of the different types of drones is presented in Table 6.1.

Table 6.1 Different types of drones and their uses

Drone Type	Pros	Cons	Use/Applications
Multi-rotor drones	Easy to control and maneuver	Its flying time is short	Aerial photography and inspection
	VTOL and hover flight	Small payload capabilities	Landing surveying
	Lower cost	Reduced wind stability	Agriculture
	Portability	Shorter flight speed	
Fixed-wing drones	Longer flight time	It requires more training before it can be used	Aerial mapping
	Ability to carry heavy payloads	No VTOL/hover	Utility inspection
	Greater stability in wind	Expensive	Surveillance
	Higher flight speeds	It requires more space and difficulty exists in landing	Agriculture
Single-rotor helicopter drones	VTOL and hover flight	More dangerous	Aerial LiDAR laser scanning
	Long endurance	More difficult to fly and requires more training	
	Heavier payload capability	Expensive	
Fixed-wing hybrid VTOL drones	VTOL and hover	More training needed	Aerial mapping
	Long-endurance flight	Expensive	Utility inspection
	Faster		Investigation
	Heavier payload capability		Agriculture

Source: JOUAV Unmanned Aircraft System (2022)

6.5.1 Multi-rotor Drones

Multi-rotor drones are the most common and globally used type. Due to its small size and excellent maneuver capabilities, it is usually the best and most preferred choice for aerial photography and for recreational and professional purposes. One of the advantages of Multi-rotor drones is their versatility which allows the installation of different types of cameras for performing diverse tasks (see Fig. 6.4).

6.5.2 Fixed-Wing Drones

Fixed-wing drones are a type of unmanned aerial vehicle (UAV) that utilize air and aerodynamic forces to remain airborne. These drones share a similar design to radio-controlled aircraft and are commonly employed for mapping large areas due

Fig. 6.4 Multi-rotor drone. (Source: JOUAV Unmanned Aircraft System)

to their impressive autonomy. By leveraging their aerodynamics and specialized design, fixed-wing drones exhibit longer endurance and faster flight speeds compared to other types of drones. However, there are some drawbacks associated with fixed-wing drones. One disadvantage is the higher acquisition cost, as they tend to be more expensive than multi-rotor drones. Additionally, fixed-wing drones require a larger landing space, similar to an airplane. Furthermore, these drones are limited to flying in a forward direction, lacking the maneuverability offered by multi-rotor drones (Smith et al., 2020)

6.5.3 Single-Rotor Helicopter Drones

Powerful and durable single-rotor drones, resembling helicopters in construction and design, possess a single rotor for power generation along with a tail for stability and directional control. Unlike other drones, they utilize gas engines instead of batteries, resulting in extended flight times. However, these drones come with certain drawbacks, including higher cost and operational complexity due to their larger blades (see Fig. 6.5) (Adams et al., 2019).

6.5.4 Fixed-Wing Hybrid VTOL Drones

Vertical take-off and landing (VTOL) drones represent the latest advancement in drone technology. These drones are essentially modified fixed-wing aircraft that have the ability to take off and land vertically (see Fig. 6.6). By combining the

Fig. 6.5 Single-rotor helicopter drone. (Source: JOUAV Unmanned Aircraft System)

Fig. 6.6 Fixed-wing hybrid VTOL drones. (Source: JOUAV Unmanned Aircraft System)

long-range capabilities and extended flight times of fixed-wing UAVs with the vertical take-off capabilities of rotary-wing devices, VTOL drones overcome the limitations of traditional fixed-wing UAVs that require expansive areas for take-off and landing. VTOL drones have been specifically designed for various applications such as mapping, surveillance, agriculture, and security purposes, offering enhanced versatility and efficiency (Jones et al., 2022).

6.6 Drone Classification by Sizes

Drones can be classified according to their sizes, which range from very small drones to large drones.

6.6.1 Nano Drones (Smallest Drones)

The nano drone is the smallest type of drone, and it can be as small as an insect. It ranges between 1 and 50 cm and can usually fit into the palm of one's hand. So, due to its compact size and unique wing design, as depicted in Fig. 6.7, they can navigate within highly very small areas without easy detection. These drones are primarily utilized by detectives to gather information pertaining to individuals and their whereabouts.

6.6.2 Medium Drones

Medium-sized drones, also known as large drones, are characterized by their larger dimensions and higher weight compared to smaller counterparts. These drones typically exceed 2 m in size and can weigh up to 200 kg (see Fig. 6.8) (Anderson,

Fig. 6.7 Nano drones. (Source: JOUAV Unmanned Aircraft System)

Fig. 6.8 Medium drones. (Source: JOUAV Unmanned Aircraft System)

2020a, b). Due to their substantial size, they usually require the assistance of two or more individuals to lift and handle them safely. Medium-sized drones are commonly utilized in professional settings, such as aerial surveying, mapping, and inspection, as well as for amateur photography purposes (Brown et al., 2019).

6.6.3 Large Drones

Large drones, primarily utilized for military operations such as surveillance and strategic maneuvers, have seen significant technological advancements. These innovations have led to their increasing adoption as replacements for traditional fighter jets, offering enhanced capabilities for swift enemy detection and combat (Smith, 2020). Moreover, large drones have also found applications in civil sectors, including drone deliveries and filmmaking, thanks to their versatility and advanced features (Johnson, 2019).

6.7 Types of Drones According to Drone Range

According to the classification scheme, this can be categorized into different ranges: very close range, close range, short range, mid-range, and long range (Brown, et al., 2019; Johnson, 2019) (Table 6.2).

Table 6.2 Drone classification according to range

Drone range	Flight distance	Flight time	Use
Very close-range drones	5 km	1 h	Recreation
Close-range drones	Up to 50 km	1–6 h	Military surveillance
			Aerial photography
Short-range drones	Up to 150 km	8–12 h	Large-scale surveillance
			Mapping and surveying
			Utility inspection
Mid-range drones	644 km	24 h	Military combat and surveillance
Long-range drones	More than 644 km	More than 24 h	Military surveillance and espionage
			Weather tracking
			Geographic mapping

Source: JOUAV Unmanned Aircraft System

6.8 Types of Drones According to Their Abilities & Equipment

6.8.1 Toy Drones

Toy drones are commonly available in the toy section of supermarkets, offering a wide range of sizes and shapes, with options for cameras or without. Primarily used for indoor activities, they can also be used outdoors depending on the specific model (see Fig. 7.2). Toy drones are typically operated using a remote control or a smartphone, providing users with various control options based on the drone's features and capabilities (Roberts, 2019).

6.8.2 Photography/Videography Drones

Photography/videography drones are specifically designed for capturing high-quality photos and videos, offering options for recording in HD, Full HD, or even 4K resolution. These drones are equipped with a camera that is mounted on a 2-axis or 3-axis nacelle, providing excellent stability for capturing clear and steady images (see Fig. 8.1). Additionally, they are often equipped with GPS technology, enabling features such as tracking and the ability to assign specific flight routes.

Fig. 6.9 Racing drones. (Source: JOUAV Unmanned Aircraft System)

6.8.3 *Racing Drones*

Racing drones, also known as FPV (First Person View) racing drones, are primarily used in competitive settings on private and tightly controlled land. These drones are specifically designed for participating in races that involve navigating through challenging obstacle courses. Pilots operate the drones at high speeds, relying on virtual reality glasses to provide a first-person perspective of the flight (see Fig. 9.1). Mastering the control of racing drones requires extensive training to develop the necessary agility and reflexes (Fig. 6.9).

6.8.4 *Professional Drones*

Professional drones, also known as commercial drones, are extensively employed in the field of aerial photography (see Fig. 6.10). These drones are equipped with advanced imaging cameras, sensors, and probes, enabling them to perform various tasks such as mapping and surveillance. The cameras integrated into professional drones are designed to capture high-quality aerial photographs from different perspectives. Moreover, these drones possess the necessary permissions and capabilities to fly over restricted or prohibited areas, including built-up regions, and can even conduct night flights, subject to proper authorization and licensing. Apart from

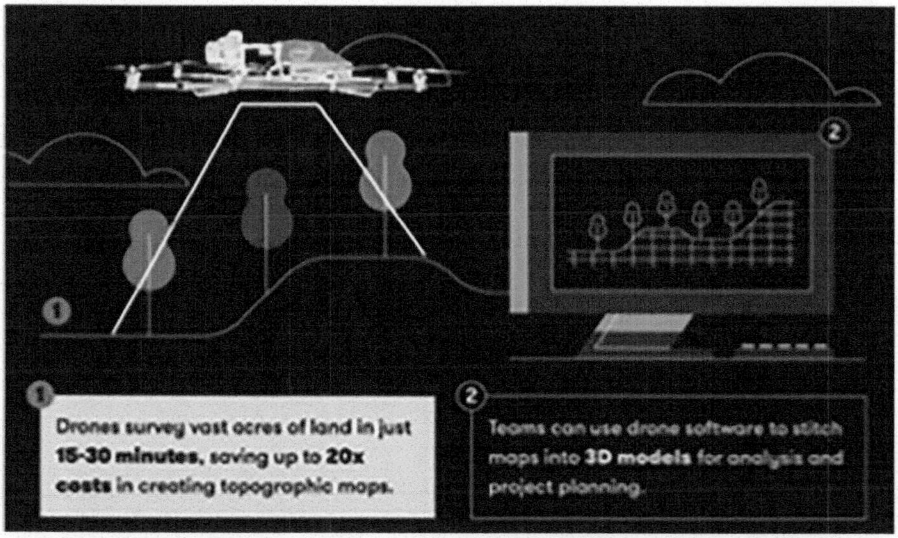

Fig. 6.10 Drone for topographic mapping and land surveys (Zitzman, 2018)

their applications in mapping and surveillance, professional drones play an important role in traffic monitoring and control, providing real-time information on the traffic conditions within urban environments.

6.8.5 Military Drones

Smith, (2021) opined that military drones serve as the foundation for many drone types and have witnessed extensive deployment in numerous conflicts. Initially employed for surveillance and intelligence gathering purposes, these drones have evolved to possess the capability of being armed, transforming into formidable weapons (see Fig. 6.11).

6.9 Application of Drones in Transportation Planning

6.9.1 Traffic Flow Analysis

Drones have proven to be valuable tools in analyzing traffic flow, enabling the assessment of different road geometries, such as roundabouts, signalized intersections, and non-signalized intersections. This analysis encompasses a range of traffic flow behaviors, including the study of shock waves, lane change patterns, and gap acceptance

Fig. 6.11 Military drones. (Source: JOUAV Unmanned Aircraft System)

analysis. Furthermore, drones can be utilized to calibrate and validate car-following models for microscopic simulation packages (Coifman et al., 2004a, b). By tracking vehicle movements at intersections and monitoring traffic congestion through video footage, UAVs provide a direct means of assessing parking lot utilization (Coifman et al., 2004a, b). In a related study, Salvo et al. (2014) utilized UAV videos to gather vehicle trajectory data and analyze gap acceptance at an urban intersection regulated by STOP signs instead of traffic signals. This analysis identified three driving behaviors: aggressive, neutral, and cautious, based on factors such as waiting time at the intersection entrance and the number of rejected gaps for drivers accepting gaps of less than 6 s (Salvo et al., 2014). These findings demonstrate the potential of unmanned aerial vehicles to provide insights into traffic behavior and control.

Additionally, Khan et al. (2018a, b) propose the use of drones to capture critical points in vehicle trajectories, allowing for the classification of different flow regimes, such as uniform motion, accelerated/decelerated motion, or stationary regimes. By analyzing these trajectories, it becomes possible to identify shockwaves generated near signalized intersections, thereby obtaining important traffic flow parameters like speed, density, and flow (Khan et al., 2018a, b) (see Fig. 6.12).

6.9.2 Highway Infrastructure Management

Another significant application of UAV technology within the transportation sector is the monitoring and management of highway physical infrastructure. This application encompasses two major areas: bridge inspection and monitoring, and pavement

Fig. 6.12 Case study of drone technology in capturing real time traffic information and monitoring. (Source: Cognitive machines, 2020)

distress recognition. Transportation infrastructure management has placed considerable emphasis on this aspect, conducting various field experiments and trials to establish protocols, methods, and algorithms. Zulkipli and Tahar (2018) conducted a study highlighting the usefulness of UAVs in mapping and road design. Specifically, they utilized small UAVs, such as Quad-Rotors and Phantom 3 Pro, for engineering surveys to develop a small-scale map for road design purposes. The images captured by the UAVs were processed using UAV Agisoft Photo Scan software, which generated the x, y, and z coordinates for the entire area of interest. The study utilized a 4 K camera, and its parameters were provided to the image processing software. Additionally, a traditional mechanism was employed in parallel to analyze the accuracy of the mapping and road design process, involving elements such as horizontal curves and superelevation.

6.9.3 Road Safety

One of the prevalent transportation challenges faced by road users is road accidents. To address this issue, the adoption of emerging technologies, such as drones, is crucial for effective control and management, ensuring the safety of motorists and pedestrians. UAVs offer various road safety applications, including detailed accident investigation, risk assessment, and overall surveillance of road networks. Risk assessment endeavors involve in-depth analysis of vehicle trajectories extracted from UAV-based videos, enabling the identification of potential conflicts and risky areas like lanes and intersections. Accident investigation studies rely on vision algorithms and systems to reconstruct accident scenes using footage and photographs captured by drones. However, challenges exist in the process of extracting information from the videos and establishing a foolproof system for the optimal functioning of drones.

In a study by Mehmood et al. (2018), the use of UAVs was compared with three other alternatives for incident management through the application of a multi-criteria decision analysis (MCDA) method. The four options evaluated were manned drones, helicopters, UAVs, and the incident commander's vehicle. The findings revealed that UAVs were more cost-effective than helicopters and manned drones, and they provided faster accident monitoring and management compared to the incident commander's vehicle. The analysis considered three crucial parameters: response time, cost, and availability of options to reach the crash site promptly and efficiently, ultimately saving lives.

6.9.4 Risk Assessment

The utilization of UAVs (Unmanned Aerial Vehicles) in transportation planning has revolutionized the extraction of vehicle trajectories, enabling microscopic-level analysis and accurate assessment of associated risks. Traditional methods of video data collection had several limitations. To address this, Zhang et al. (2013) introduced an algorithm designed to detect collisions in hazardous situations by examining vehicle trajectories captured from UAV-based traffic video data. Building on this, Gu et al. (2019) presented a novel analysis framework that also utilizes UAV-based video data to extract vehicle trajectories. Their research concentrated on assessing crash risks associated with mandatory lane changes in freeway merging zones. They proposed a method to evaluate crash risk between merging vehicles by developing a merging behavior model that includes the safety metric of time-to-collision.

Kim and Chervonenkis (2015) investigated on-board automatic detection of emergency and abnormal traffic situations using UAV artificial vision systems. They proposed a system to detect stationary and moving vehicles based on UAV video data while simultaneously identifying and classifying abnormal situations to assist operators in making informed decisions. However, their algorithms were limited to recognizing only a few abnormal situations. Sharma et al. (2017) conducted advanced research on the use of UAVs for risk assessment in transportation. They employed a multi-UAV coordinated vehicular network to analyze driving behavior and improve road safety. Their motivation stemmed from the limitations of infrastructure-based vehicular ad hoc networks (VANETS), which are constrained by traffic and have limited scope. By using multiple UAVs, they created an independent system that enhanced communication and achieved more accurate vehicle tracking. The study concluded that their proposed system outperformed existing state-of-the-art approaches. Furthermore, Iglesias et al. (2019) employed UAV video data obtained through double grid UAV flights to assess appropriate sight distances. They processed a dense point cloud using Motion Multi-view Stereo software, classifying it to generate a terrain model and a 3D object model. Sight distances were then calculated based on the resulting environmental model. The study emphasized that with increasing UAV usage, abundant data would be available, leading to improved risk assessment methods in the future.

6.9.5 Accident Investigation

Accident investigations have greatly benefited from the use of drones (Unmanned Aerial Vehicles). Photogrammetry-based traffic incident scene generation for accident investigations has been a topic of study since the 1990s. However, this field faced several challenges that have been overcome with the advancement of technologies, particularly with the introduction of drones. Outay et al. (2020) highlighted that drones have played a crucial role in accident investigation by enabling accurate examination and analysis of accident causes at a minimal cost compared to manual and conventional methods (Figs. 6.13, 6.14 and 6.15).

Fig. 6.13 Accident scene mapping with a drone (PhotoModeler, 2020)

Fig. 6.14 Accident scene mapping with a drone (PhotoModeler, 2020)

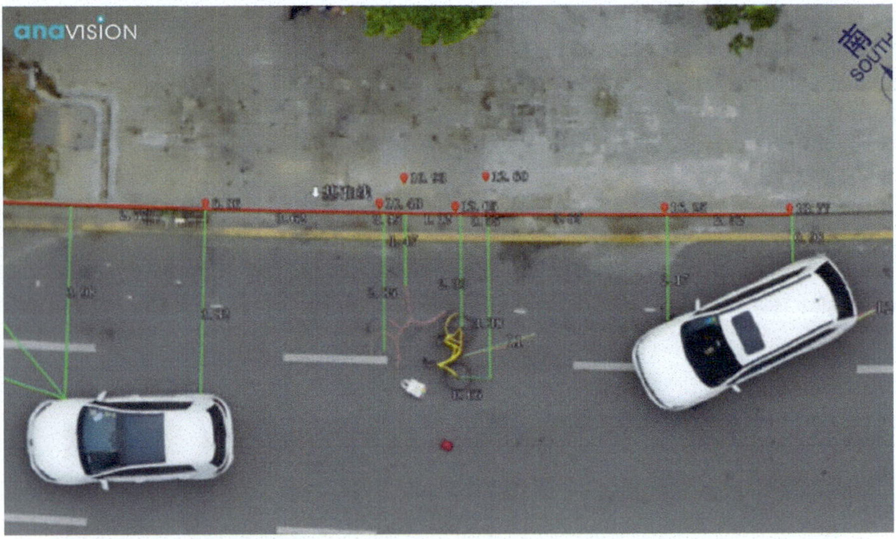

Fig. 6.15 Accident investigation with aerial photography (Anavision 2019)

6.10 Required Skills Needed by Drone Operators for Effective Transportation Planning

In recent years, the significance of drone technology in transportation planning cannot be overstated, given its numerous benefits. However, to ensure the effective utilization of this technology, particularly in the transportation sector, operators need to possess specific skills before being licensed or allowed to fly drones in built-up areas. This requirement is crucial to guarantee the safety of both the operators and the society as a whole.

- **Data Analytics Skills**: Drones, also known as unmanned aerial vehicles (UAVs), are technological devices that collect, store, and transmit data. Their usage has expanded beyond video capture, finding application in various domains, including traffic monitoring and predicting future traffic growth (Adepoju et al., 2022). Modern UAVs are designed to generate large datasets, highlighting the importance of proficiency in data analytics. In the realm of transport planning, the ability to analyze drone data and extract valuable insights is crucial. Data analytics skills enable the processing of drone data, facilitating data accessibility for project teams and informed decision-making. Moreover, these skills play a vital role in addressing any potential issues related to the generated drone data.
- **Programming Skills**: In light of the rapid advancement of technology relying on computer programs, there is a growing worldwide need for programming proficiency. This is particularly evident in the field of drone operation, where programming skills are essential for effectively managing, processing, and analyzing the data gathered by drones. As UAV systems become increasingly autonomous, integrating diverse autopilot modes, the importance of advanced programming skills becomes even more pronounced. These skills enable operators to activate flight assistance functionalities like homing, leveling, automatic takeoff, and landing methods, enhancing the overall performance and control of advanced UAV systems.
- **Machine Learning Skills**: Drones are recognized as a crucial emerging technology that plays a vital role in successful transportation and urban planning. In this context, machine learning has become increasingly important in automating the analysis of aerial data. Proficiency in machine learning enables the extraction of valuable information from data, including object detection and the transformation of extensive datasets into easily consumable reports. Moreover, machine learning has addressed key challenges in unmanned aerial systems, particularly in achieving autonomous flight capabilities. By leveraging machine learning, the ability to identify and predict patterns in data and develop software for autonomous flight can be significantly enhanced.

To summarize, drone technology has emerged as a pivotal technology in effective transportation planning. Unmanned aerial vehicles (UAVs) have become indispensable, bringing digital transformation to the built environment. By integrating various components such as sensors, GPS navigation units, laser scanners, and high-resolution cameras, drone capabilities have been greatly enhanced for commercial applications worldwide. UAVs offer numerous benefits, including enriching databases to predict congestion phenomena, improving response times in challenging scenarios, and preventing accidents. With their near-real-time data collection and monitoring capabilities, drones enable the coverage of large areas that would otherwise be impractical with single inductive or pneumatic sensors.

6.11 Application of Geographic Information System (GIS) in Traffic Management

The current systems of manual traffic control or predefined times for traffic light changes have proven to be inefficient. The objective of traffic congestion control and management systems is to clear accumulating traffic as quickly as possible and prioritize the passage of emergency vehicles (Chandana et al., 2017a, b). It is evident that when one lane has fewer vehicles while another has more, the green signal is often displayed for the same duration, resulting in wasted time for empty roads. This emphasizes the fact that transportation decisions have a significant impact on human life and our urban spaces (Miller & Shaw, 2001a, b). To address some of these urban transport problems, it is essential to adopt advanced technologies and software, such as Geographic Information System (GIS), which is less prone to human errors.

GIS, or Geographic Information Systems, has been hailed as one of the most significant advancements in geography since its inception in the 1970s. Its value in the transportation sector has been widely recognized, leading many transportation agencies to embrace GIS and Geospatial Information Systems for Transportation (GIS-T) as indispensable tools. One of the key advantages offered by GIS, as highlighted by Olba Albu (2006), is its ability to manage spatial data in layers and overlay them for spatial analysis. For example, it enables the integration of a roads layer with a land use layer, facilitating buffer analysis to assess land uses within a specified distance from the road. Over the past few decades, GIS has proven highly effective in providing spatial information for transportation planning, continuously expanding its capabilities and offering a diverse range of data management and analysis tools. In the early 1990s, GIS introduced specific tools for the management of linear transportation data, which have been widely adopted by transportation organizations. These tools allow transit agencies to georeference their bus routes, stops, time points, and other relevant features to a digital street centerline file, ensuring the synchronization of these data (Sutton et al., 2004). The deployment of GIS has also attracted the attention of transportation software vendors, who offer scheduling, vehicle tracking, and trip itinerary planning programs. Consequently, this technology has delivered substantial benefits in addressing transportation-related challenges, particularly in urban areas experiencing rapid increases in car ownership.

According to Bolstad (2016), GIS is a computer-based system that facilitates the collection, storage, analysis, output, and distribution of spatial data and information. Its application extends to various fields and can contribute to improved well-being, economic prosperity, and enhanced safety. In transportation, GIS and spatial analysis focus on quantitatively locating important features and examining their associated properties and attributes. Like in other domains, GIS plays a vital role in achieving excellence in transportation by assisting in the planning, monitoring, and management of complex transportation systems. Through GIS, a country's transportation system can be optimized by enhancing capacity and making informed investments in transportation infrastructure.

The advantages of GIS in data documentation and processing, as noted by Al-Ramadan and Aina (2004), include:

- Quick updating of information
- Automated cartography
- Integration of spatial and attribute data
- Spatial analysis
- Production of maps at different scales
- Visualization

Furthermore, Al-Ramadan (2002) highlighted various urban applications of GIS, such as:

- Permit issuing and tracking
- Municipal Facilities Management
- Zoning and Subdivision Plan Review
- Urban Land Use Planning
- Land Records Management
- Emergency Vehicle Routing and Dispatching
- Inventory of Utilities
- Inventory of vacant Land parcels
- Urban Growth Management
- Transportation Analysis and Planning
- Optimum Site Selection for schools, hospitals, and commercial centers.

In conclusion, GIS plays a vital role in shaping the urban and transportation systems of a country through its diverse functions. It provides the means to optimize transportation planning, enhance data management and analysis, and improve decision-making processes. By leveraging GIS technology, transportation agencies can effectively address urban transport.

6.12 Components of GIS

Several components such as hardware, software, data, and human features make which ensures its proper functioning.

6.12.1 Hardware

GIS heavily relies on computer systems as the fundamental hardware component (ESRI, n.d.-a). For most GIS applications, a high-speed computer with ample data storage capacity and a high-quality, large display is essential (ESRI, n.d.-a). This requirement stems from the nature of spatial analyses, which often encompass extensive areas and involve high spatial resolutions. Due to the substantial amount

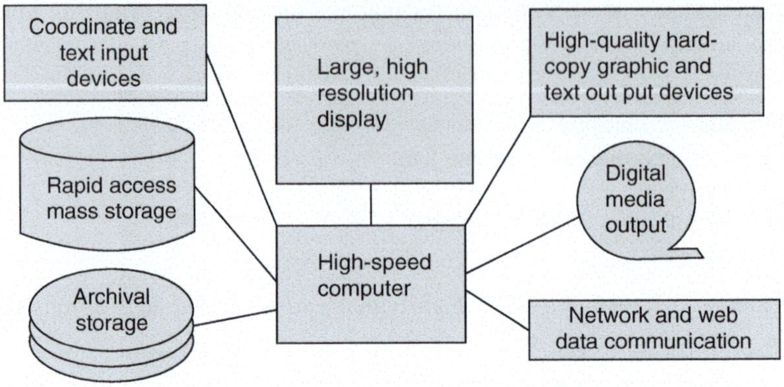

Fig. 6.16 Specialized GIS Hardware Components (Bolstad, 2016)

of data involved in GIS, specialized equipment becomes necessary to facilitate efficient data entry tasks (ESRI, n.d.-a). Accurately defining the shape and location of geographic features, such as roads, rivers, and parcels, is crucial in GIS (ESRI, n.d.-a). This necessitates the utilization of large volumes of data, which must be entered into the system with precision (ESRI, n.d.-a). To streamline and enhance the data entry process, specialized equipment is required (ESRI, n.d.-a) (Fig. 6.16).

6.12.2 GIS Software

To ensure the effective functioning of GIS, specialized software is essential for managing, analyzing, displaying, and disseminating spatial information. This software enables GIS to handle tasks such as collecting and manipulating coordinates, as well as working with qualitative and quantitative data related to the nonspatial attributes of geographic features. It provides the necessary tools for viewing, editing, and extracting the required information from the data. Various GIS software packages are available to fulfill these requirements. Some commonly used tools include ArcGIS, QGIS, and MapInfo.

6.12.2.1 ArcGIS

ArcGIS, considered one of the most widely used GIS software, offers an extensive range of geoprocessing procedures, encompassing tasks from data entry and analysis to diverse forms of data output. It is specifically designed to support various data formats, encompassing a wide array of data types and structures. With its comprehensive functionality, ArcGIS enables the execution of numerous operations on spatial data, presenting users with thousands of possibilities. The utilization of ArcGIS has proven to be advantageous, providing flexibility in modeling geographic features.

6.12.2.2 Idrisi

Idrisi, a GIS software developed by the Graduate School of Geography at Clark University, Massachusetts, stands out among other GIS software packages due to its distinctive combination of image processing and GIS functions. Recognizing the value of image data as a valuable source of information in GIS, Idrisi places emphasis on tasks related to image data collection, manipulation, and output. In addition to its image processing capabilities, Idrisi offers a comprehensive suite of spatial data analysis and display functions. This software package is particularly beneficial for beginners as it not only facilitates the use of GIS but also enables users to develop their own spatial analysis functions, thereby fostering learning and skill development.

6.12.2.3 AUTOCAD MAP 3D

AUTOCAD, developed by Autodesk, Incorporated of San Rafael, California, is a widely popular computer drafting and design software package (Autodesk, n.d.). Originally designed as an engineering drawing and printing tool, AUTOCAD has expanded its utility to various fields, including surveying, where it is used for managing property boundary coordinates. In addition, AUTOCAD MAP 3D, a specialized version of AUTOCAD, offers advanced spatial data management and analysis capabilities (Autodesk, n.d.). Its comprehensive set of features includes enhanced analytical tools that enable users to perform in-depth spatial analysis tasks (Autodesk, n.d.).

6.12.2.4 Maptitude

Maptitude is a powerful GIS software that specializes in spatial analysis within a business setting (Caliper Corporation, n.d.). It offers advanced capabilities for integrating demographic attributes, such as income and occupation, with location and data allocation. This makes it particularly useful for transportation and logistics businesses, as well as other functions across various industries (Caliper Corporation, n.d.). Maptitude stands out for its ability to seamlessly integrate diverse datasets, including census, health, and infrastructure data (Caliper Corporation, n.d.). Additionally, it provides utilities for server and Web-based data export, map creation, and display, enhancing accessibility and visualization of spatial information (Caliper Corporation, n.d.).

6.13 GIS in Transportation Planning

The use of Geographic Information System (GIS) in transportation planning agencies, particularly among metropolitan transportation organizations, is increasingly widespread. GIS integrates hardware, software, and data to capture, manage, analyze, and display geographically referenced information (Gupta et al., 2003a, b). In many developed countries, the maintenance of highways has become a critical issue, prompting authorities to adopt GIS for highways and transport management due to its cost-effectiveness and growing familiarity (Gupta et al., 2003a, b). GIS offers transportation planners a powerful tool for storing and analyzing various data related to population densities, land uses, and travel behavior, among other factors (Gupta et al., 2003a, b). This information obtained from GIS has been instrumental in the development of transportation policies and planning in both the public and private sectors (Gupta et al., 2003a, b).

The applications of GIS in transportation are extensive, including highway maintenance, traffic modeling, accident analysis, route planning, and environmental assessment of road schemes (Harris & Batty, 1993). A well-structured road network is a fundamental requirement for transportation, and GIS provides additional information on general topography, land cover, and land use, which are crucial for evaluating the impact of construction projects (Harris & Batty, 1993). GIS represents the world through map visualization, utilizing longitudes, latitudes, and other projection systems, with a hierarchical structure of graphical objects (Harris & Batty, 1993). Despite its numerous benefits, the availability of appropriate data for GIS remains an ongoing challenge (Harris & Batty, 1993). Ensuring the quality and accessibility of data sources is essential for maximizing the potential of GIS in transportation planning.

6.14 Roles of GIS in Transportation Planning

In many modern societies, roads are the key medium for transport through the numerous functions they perform. However, the trends in urbanization and population have increased the number of vehicles on our roads thus causing so many transportation problems. As a result of this, stakeholders and government in transportation sectors have explored several solutions to minimize some of these transport problems which have plagued faced society. In this regard, application of GIS in road traffic management renders many useful functions for assuring an efficient flow of traffic in the transport systems of a country.

6.14.1 GIS in Highway Management

Highway maintenance management is a pressing concern that can be effectively addressed through the utilization of Geographic Information System (GIS). The adoption of GIS for highways and transport management is widespread among authorities in developed countries (Shimonti, 2018a, b, c). One of the primary advantages of GIS is its ability to visualize data, often accomplished through the creation of maps. This visualization of real-time data empowers transport planners

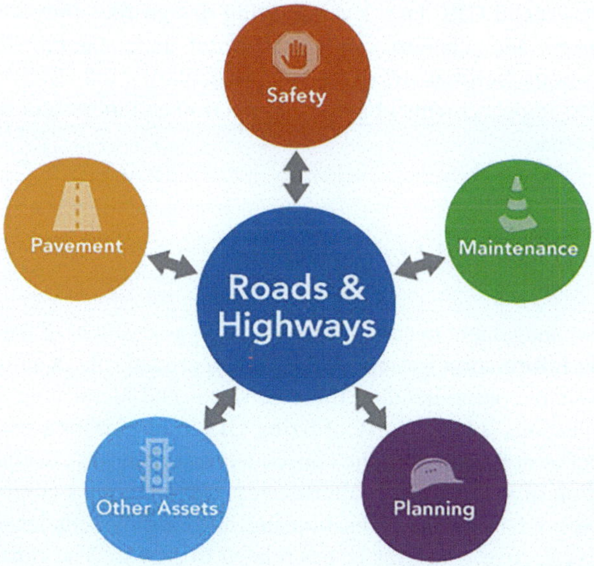

Fig. 6.17 Chains of activities in highway maintenance

to easily identify potential issues and enables more efficient and cost-effective solutions compared to traditional methods. The information derived from GIS can be communicated to decision-makers who can incorporate it into highway maintenance and management strategies (Shimonti, 2018a, b, c).

To illustrate, ESRI's mobile GIS technology offers a valuable solution for transportation maintenance and inspection crews (see Fig. 6.17). This technology allows them to streamline data collection and updates from remote locations, saving time and improving efficiency (ESRI, n.d.-b). By utilizing GPS and GIS-enabled devices, field crews can accurately gather information in the field and seamlessly update central databases in real-time, eliminating the need for manual data transfer (ESRI, n.d.-b). Additionally, through GPS and Esri technology, officials can monitor resources and assets in real-time, facilitating rapid responses to critical events that require immediate attention (ESRI, n.d.-b). Furthermore, the ability to track the location of work or inspection crews on a map enables efficient resource management and allows for quick redeployment as needed (ESRI, n.d.-b).

6.14.2 GIS and Planning or Locational Design

Transportation planners have seamlessly incorporated Geographic Information System (GIS) into their design processes by incorporating imagery, elevation, and environmental data. This integration allows planners to leverage familiar software while accessing crucial GIS data. By importing design files into a GIS platform, transport engineers and planners can gain a better understanding of the optimal placement of various facilities to ensure complementarity and efficiency. This integration has led to improvements in material selection and more accurate estimation of total project costs.

6.14.3 Environmental Management

Environmental management necessitates a thorough assessment of the environment, and Geographic Information System (GIS) plays a crucial role in aiding transportation professionals in comprehending these concerns and identifying the most environmentally conscious solutions. By utilizing GIS, transportation practitioners can develop a deeper understanding of the consequences of land-use decisions, facilitating the evaluation of smart-growth alternatives. GIS integrates environmental factors with analyses of land-use patterns, housing, and employment densities, enabling communities to effectively address issues related to growth. The capability to visualize different alternatives on a shared platform facilitates consensus-building among stakeholders with differing perspectives, particularly in matters involving environmental sensitivity.

6.14.4 GIS and Traffic Modelling

The utilization of traffic modelling incorporating GIS data provides valuable insights into the movement patterns of individuals and communities, offering transport planners enhanced information about both people and their surrounding environment (Aderibigbe & Gumbo, 2022). This enables more accurate predictions of future growth and facilitates better planning strategies. GIS data can be leveraged to develop functional road models for large-scale traffic simulation by representing road networks worldwide as polylines with attributes. These roadmaps, derived from GIS databases, can automatically generate geometrically accurate and topologically consistent 3D models of expansive road networks. Such models find applications in real-time traffic simulation, interactive visualization of virtual environments, and autonomous vehicle navigation. Additionally, the resulting model representations offer valuable road features for conducting traffic simulations (Aderibigbe & Gumbo, 2022).

The adoption of GIS in traffic modelling yields numerous benefits, including:

- Enhanced ease of movement within transportation systems
- Reduction in waiting times and delays
- Improved mobility of individuals
- Increased roadway safety and personal protection

- Efficient transport planning and design
- Reliable and safe operation of transportation networks

By incorporating GIS technology, transport planners can optimize traffic flow, enhance safety measures, and ensure efficient transportation operations (Aderibigbe & Gumbo, 2022).

6.14.5 *Safety Management*

Road accidents are a significant negative consequence of traffic congestion (Aderibigbe & Ogunmola, 2020). Therefore, it is essential to have accurate records of accident locations to improve safety for various road users, including motorists, freight carriers, railways, and pedestrians. GIS maps provide a valuable tool for displaying crash records along with spatial analyses of congestion, construction zones, and weather conditions. This visual representation helps reveal crucial information that might be overlooked when examining simple tabular data. By combining spatial analysis with statistical and business intelligence tools, it becomes possible to identify the root causes of accidents and develop effective countermeasures. Furthermore, GIS enables the identification of trends, such as increased oversized vehicle traffic, permit violations, and general commercial traffic route information.

6.14.6 *GIS and Accident Minimization*

Traffic accidents pose significant challenges to society, and their implications have political, economic, and social consequences within a country (Shimonti, 2018a, b, c). To address this issue, it is crucial to manage traffic accident information systems that can analyze and provide access to spatial and descriptive data. GIS offers powerful spatial analytics capabilities that enable authorities to discover patterns and gain insights into travel behaviors and accident analysis. Upadhyay (2018a, b) emphasized the role of GIS in accident reduction by providing relevant data and information about accident-prone areas. This information, presented in a geographical format, highlights dangerous locations and intersections with a high likelihood of accidents. It serves as a guide for road users to exercise caution when approaching these areas.

In summary, the application of GIS in capturing spatial information has proven to be immensely beneficial for urban transportation planning. This chapter has explored the role of emerging technologies, such as Unmanned Aerial Vehicles (UAVs) or Drones, alongside Geographic Information Systems (GIS), in sustainable planning and smart city development. These technologies have digitally transformed the built environment and have become indispensable in urban transport planning. The combination of drones with various tools like sensors, GPS navigation units, laser scanners, and high-resolution cameras has significantly enhanced their capabilities and efficiency for urban planning and management. The use of these technologies has greatly contributed to transportation planning by reducing accidents, enabling real-time monitoring of traffic flow, managing public transport, and enhancing decision-making processes.

6.15 The Role of Machine Learning Technology in Traffic Management

The world is rapidly transforming with exponential population growth and technological advancement which has been occurring swiftly. Based on this, urban and transport planners have long sought sustainable practices to address traffic-related issues such as congestion, accidents, and delays in cities worldwide. However, managing traffic becomes increasingly challenging as the global population grows. To enhance traffic control efficiency, modern technical equipment and training for traffic officers are essential. Machine learning, which involves creating algorithms that can learn from data and improve over time, offers promising solutions (Bisen et al., 2023). According to Mbowa et al. (2021), the world is currently undergoing the fourth industrial revolution, characterized by automation and digitization, resulting in increased use of automated devices and reduced reliance on human labor. Hence, a need for more advanced technology such as machine learning and intelligent based transportation system to manage and predict traffic growth.

According to Patil (2022), traffic prediction involves utilizing machine learning models to forecast traffic patterns within a specific area. In recent years, machine learning has gained traction, with numerous cities leveraging it to manage traffic and alleviate congestion. Deep learning models have emerged as particularly effective tools for traffic prediction due to their capacity to analyze extensive datasets and generate precise forecasts. These models employ neural networks to examine large datasets, making them adept at discerning intricate patterns that may elude human analysts. Consequently, traffic managers can anticipate congestion and adjust traffic signal timings proactively. One notable deep learning model applicable to traffic prediction is the Convolutional Neural Network (CNN), traditionally employed for image recognition but adaptable to traffic prediction tasks. CNNs process sequences of traffic images or videos through convolutional layers to discern patterns and make predictions accordingly. Similarly, the Recurrent Neural Network (RNN) is adept at handling sequential data, such as traffic data. By incorporating recurrent layers, RNNs can analyze past traffic patterns to inform future predictions (Ball et al., 2017).

Utilizing deep learning models for traffic prediction offers several advantages. Machine learning's rapid and accurate evaluation of vast datasets enables transport planners to make well-informed decisions regarding traffic flow and congestion

(Guo et al., 2016). Additionally, deep learning models exhibit adaptability to changing traffic conditions, as they can be continually trained on new data, ensuring increasingly accurate predictions over time. Furthermore, the scalability of deep learning models allows for their application across a spectrum of traffic management scenarios, from individual intersections to entire urban areas, facilitating tailored solutions for diverse needs.

6.16 Importance of Machine Learning in Transport Planning

6.16.1 Applications of Deep Learning in Object Detection and Recognition

The ability to swiftly detect and identify objects in real-time holds significant importance across various domains, including traffic systems and self-driving cars. Deep learning algorithms, a subset of machine learning, have demonstrated remarkable potential in enhancing the efficiency of object detection and recognition systems. Object detection, a computer vision task, involves pinpointing and categorizing objects within images or videos, aiming to identify their presence and location within frames. One widely recognized deep learning algorithm for object detection is the You Only Look Once (YOLO) algorithm. YOLO divides an image into a grid of cells and predicts bounding boxes and class probabilities for each cell. Renowned for its speed and accuracy, the YOLO algorithm is well-suited for real-time applications such as self-driving cars and surveillance systems (Patil, 2022). Machine learning can complement other traffic management techniques, such as dynamic traffic signal control systems or traffic rerouting algorithms. Integrating these diverse approaches enables traffic managers to develop a comprehensive traffic management system that effectively reduces congestion and enhances traffic flow.

6.16.2 Applications of Deep Learning in Autonomous Vehicles

The rise of autonomous vehicles marks a significant shift in transportation, introducing vehicles capable of operating without human intervention. Deep learning stands out as a pivotal technology driving the advancement of autonomous vehicles, offering considerable potential in enhancing their safety, efficiency, and reliability. As a subset of machine learning, deep learning utilizes intricate neural networks to detect patterns and glean insights from data. In the realm of autonomous vehicles, deep learning models undergo training on extensive datasets comprising images, video feeds, and sensor data, enabling vehicles to perceive and react to their surroundings effectively. This iterative process empowers autonomous vehicles to make informed real-time decisions, such as adapting their speed and trajectory to navigate obstacles or changing road conditions.

6.16.3 Applications of Machine Learning in Traffic Flow Optimization

Urban planning places significant emphasis on optimizing traffic flow to enhance efficiency and alleviate congestion. Deep learning models have emerged as promising tools in this domain, leveraging real-time traffic data to adjust signal timings, forecast traffic patterns, and redirect traffic routes. Moreover, these models offer a pathway to sustainable traffic flow optimization by mitigating congestion and reducing greenhouse gas emissions from vehicles, particularly crucial in urban areas plagued by air pollution and climate change concerns. Another potential application of deep learning models lies in emergency response scenarios. During natural disasters or other emergencies, these models can facilitate swift and safe rerouting of traffic, ensuring expedited access for emergency vehicles. This capability proves invaluable in saving lives and minimizing post-disaster damage. One of the primary advantages of employing deep learning models for traffic flow optimization is their adeptness at processing vast datasets and discerning patterns over time. This enables accurate predictions of traffic patterns, facilitating timely adjustments to signal timings. Additionally, these models demonstrate adaptability to evolving traffic dynamics, learning and accommodating changes stemming from shifts in population density or urban development. Thus, deep learning models offer a dynamic and efficient approach to optimizing traffic flow in urban environments.

6.16.4 Applications of Deep Learning in Parking Management

The integration of deep learning technology into parking management systems has revolutionized the efficiency of parking operations. Traditional methods relied on manual monitoring, resulting in time-consuming processes and frustration for drivers. In contrast, deep learning-based systems utilize computer vision and machine learning algorithms to detect parking spaces in real-time, significantly enhancing efficiency. By swiftly identifying available parking spaces, deep learning-powered systems reduce the time spent by drivers searching for spots, thereby lowering fuel consumption and improving air quality. Additionally, these systems guide drivers to vacant parking spots efficiently, enhancing the overall parking experience. Moreover, they contribute to reducing parking violations by monitoring parking spaces and identifying illegally parked vehicles. This enables parking authorities to issue tickets promptly and enforce regulations effectively. Furthermore, deep learning algorithms can recognize license plates and identify vehicles that have exceeded their parking time limits, deterring drivers from violating parking regulations and mitigating traffic congestion in urban areas. The environmental benefits are notable, with reduced fuel consumption leading to lower emissions and the efficient use of parking spaces potentially minimizing the need for additional parking structures, thus preserving urban green spaces. Moreover, deep learning technology can manage parking spaces in high pedestrian traffic areas, reducing the risk of accidents caused by drivers searching for parking spots. While deep learning algorithms offer significant potential in anticipating traffic patterns and enhancing urban traffic management, they do have limitations, including the requirement for extensive historical data and ongoing maintenance and upgrades. Nonetheless, leveraging deep learning models for traffic forecasting presents a viable approach to reducing traffic congestion and improving transportation efficiency.

References

Adams, J., Smith, R., & Thompson, L. (2019). Single-rotor drones: Construction, design, and operational considerations. *Journal of Unmanned Aerial Systems, 15*(2), 45–58.

Adepoju, O., Aigbavboa, C., Nwulu, N., & Onyia, M. (2022). *Re-skilling human resources for construction 4.0*. Springer Tracts in Civil Engineering. https://doi.org/10.1007/978-3-030-85973-2_1

Aderibigbe, O. O., & Gumbo, T. (2022). Modelling the variation in the mobility pattern of households in the urban and rural areas of Nigeria. *Scientific Journal of Silesian University of Technology. Series Transport, 117*, 5–21. ISSN: 0209-3324. https://doi.org/10.20858/sjsutst.2022.117.1

Aderibigbe, O. O., & Ogunmola, J. (2020). Trends of road traffic crashes in Akure, Nigeria. *Nigerian Journal of Logistics and Transport, 2*, 219–228.

Al-Ramadan, B. (2002). *Introduction to geographic information systems technology and its applications, review 2002: College of Environmental Design* (pp. 113–120). KFUPM.

Al-Ramadan, B., & Aina, Y. (2004). *GIS applications in optimum site selection for tourist sites: Texas State as a case study*.

Anavision. (2019). *Accident investigation with aerial photography*. https://anavision.com/blog/traffic-accident-investigation. Accessed 20 Jan 2023.

Anderson, C. (2020a). *The drones book: Master the art of aerial photography and cinematography* (2nd ed.). Imagine Publishing.

Anderson, J. (2020b). Advanced features and capabilities of photography/videography drones. *Journal of Aerial Imaging, 25*(2), 45–58.

Autodesk. (n.d.). *AutoCAD Map 3D*. Retrieved from https://www.autodesk.com/products/autocad-map-3d/overview

Azza, A., Koubaa, A., Mohamed, N. A., Ibrahim, H., Ibrahim, Z., Kazim, M., Ammar, A., Benjdira, B., Khamis, A., & Hameed, I. (2021a). Drone deep reinforcement learning: A review. *Electronics, 10*, 999.

Azza, M. A., Kamil, S., Mohd Shafie, A., Anuar, M. S., & Omar, K. (2021b). Applications of drones in transportation: A review. *International Journal of Advanced Science and Technology, 30*(12), 3080–3093.

Ball, J. E., Anderson, D. T., & Chan, C. S., Sr. (2017). Comprehensive survey of deep learning in remote sensing: Theories, tools, and challenges for the community. *Journal of Applied Remote Sensing, 11*(4), 042609.

Bisen, K., Shahare, R., Wasnik, K., & Jaipurkar, P. P. (2023). Machine learning-based intelligent traffic system. *International Research Journal of Modernization in Engineering Technology and Science*. https://doi.org/10.56726/IRJMETS35340

Bolstad, P. (2016) *GIS Fundamentals: A First Text on Geographic Information Systems*. Eider Press, Minnesota.

Brahimi, F., Boufaida, M., Brahimi, M., & Boussaid, O. (2020a). UAV-based traffic monitoring system for smart cities. In *Proceedings of the international conference on intelligent systems design and applications* (pp. 94–103). Springer.

Brahimi, M., Karatzas, S., Theuriot, J., & Christoforou, Z. (2020b). Drones for traffic flow analysis of urban roundabouts. *International Journal of Traffic and Transportation Engineering, 9*(3), 62–71. https://doi.org/10.5923/j.ijtte.20200903.02

Brown, D., Schwer, D., & McLin, B. (2019). *Drone operator's logbook: An essential tool for professional and recreational drone pilots*. CreateSpace Independent Publishing Platform.

Butila, E. V., & Boboc, R. G. (2022a). Urban traffic monitoring and analysis using unmanned aerial vehicles (uavs): A systematic literature review. *Remote Sensing, 2022*(14), 620. https://doi.org/10.3390/rs14030620

Butila, M., & Boboc, C. (2022b). Smart city drones: Current applications and future trends. *Journal of Ambient Intelligence and Humanized Computing, 13*(3), 3159–3174.

Caliper Corporation. (n.d.). *Maptitude overview*. Retrieved from https://www.caliper.com/maptovu.htm

Chandana, S., Sundaram, S., D'sa, A., Swamy, P., & Navya, H. (2017a). Intelligent traffic control system for emergency vehicle clearance using image processing. In *2017 IEEE international conference on computational intelligence and computing research (ICCIC)* (pp. 1–4). IEEE.

Chandana, K. K., Sundaram, S. M., D'sa, C., Swamy, M. N., & Navya, K. (2017b). A smart traffic management system for congestion control and warnings using Internet of Things (IoT). *Saudi Journal of Engineering and Technology, 2*(5), 192–196.

Cognitive Machines. (2020). *Real time traffic video analytics using drone*. https://cognitivemachines.in/portfolios/traffic-video-analytics. Accessed Dec 2022.

Coifman, B., Evers, M., & Al-Begain, K. (2004a). Unmanned aerial vehicles for traffic monitoring. In *Proceedings of the international conference on intelligent transportation systems* (pp. 400–405). IEEE.

Coifman, B., McCord, M., Mishalani, R. B., & Redmill, K. (2004b). Surface transportation surveillance from unmanned aerial vehicles. In *Proceeding of the 83rd annual meeting of the Transportation Research Board*. Transportation Research Board.

ESRI. (n.d.-a). *ArcGIS*. Retrieved from https://www.esri.com/en-us/arcgis

ESRI. (n.d.-b). *Mobile GIS for transportation maintenance and inspection*. Retrieved from https://www.esri.com/en-us/industries/transportation/operations/mobile-gis-for-transportation-maintenance-and-inspection

Giordan, D., Adams, M. S., Aicardi, I., Alicandro, M., Paolo, A., Marco, B., et al. (2020). The use of unmanned aerial vehicles (UAVs) for engineering geology applications. *Bulletin of Engineering Geology and the Environment, 79*, 3437–3481. https://doi.org/10.1007/s10064-020-01766-2

Gu, X., Abdel-Aty, M., Xiang, Q., Cai, Q., & Yuan, J. (2019). Utilizing UAV video data for in-depth analysis of drivers' crash risk at interchange merging areas. *Accident Analysis & Prevention, 2019*(123), 159–169.

Guo, Y., Liu, Y., Oerlemans, A., Lao, S., Wu, S., & Lew, M. S. (2016). Deep learning for visual understanding: A review. *Neurocomputing, 187*, 27–48.

Gupta, P., Jain, N., Sikdar, P.K. & Kumar, K. (2003a). Geographical information system in transportation planning. Map Asia Conference.

Gupta, P., Parida, M., & Mishra, S. (2003b). GIS applications in transport planning: A review. *Journal of Indian Roads Congress, 64*(1), 85–105.

Gupta, A., Afrin, T., Scully, E., & Yodo, N. (2021a). Advances of UAVs toward future transportation: The state-of-the-art, challenges, and opportunities. *Future Transportation, 1*, 19.

Gupta, S., Bopche, P., & Jagtap, R. (2021b). Application of drones in civil engineering: A review. *International Journal of Advanced Engineering Research and Science, 8*(4), 39–46.

Harris, B., & Batty, M. (1993). The development of geographic information systems for transport planning. *Transportation Research Part A: Policy and Practice, 27*(6), 447–461.

Hassanalian, M., & Abdelkefi, A. (2017). Classifications, applications, and design challenges of drones: A review. *Progress in Aerospace Sciences, 91*, 99–131. Retrieved from https://www.researchgate.net/profile/Mostafa_Hassanalian/publication/316673697_Classifications_applications_and_design_challenges_of_drones_A_review/links/59ca8917aca272bb05078e84/Classifications-applications-and-design-challenges-of-drones-A-review.pdf

Iglesias, L., Santos-Berbel, D., Pascual, V., & Castro, M. (2019). Using small unmanned aerial vehicle in 3D modeling of highways with tree-covered roadsides to estimate sight distance. *Remote Sensing, 11*(22), 2625.

Johnson, R. (2019). The versatility of large drones in civil sectors. *International Journal of Drone Applications, 12*(2), 78–92.

Jones, A., Smith, B., & Johnson, C. (2022). Vertical take-off and landing (VTOL) drones: Advancements and applications. *International Journal of Unmanned Systems, 25*(3), 78–92.

JOUAV Unmanned Aircraft System (2022). Different types of drones and uses. Retrieved from https://www.jouav.com/blog/drone-types.html. Accessed Jan 2023.

Karan, E., Christmann, C., Gheisari, M., Irizarry, J., & Johnson, E. (2014a). A comprehensive matrix of unmanned aerial systems requirements for potential applications within a Department of Transportation. *Construction Research Congress, 2014*, 964–973.

Karan, M., Rahmouni, A., Zou, Y., Pappas, I., & Liu, H. X. (2014b). Unmanned aerial vehicle applications for transportation infrastructure monitoring and management. In *Proceedings of the international conference on computing in civil and building engineering* (pp. 879–886). ASCE.

Karla, M., Raut, S. S., & Ingole, N. V. (2021). A review on intelligent transportation systems using unmanned aerial vehicles. In *Proceedings of the international conference on computing, communication, and networking* (pp. 195–200). Springer.

Kazagli, E., Xekalakis, P., Tsogas, M., & Kantartzis, A. (2022). Integration of unmanned aerial vehicles in traffic management: Current status and challenges. *IEEE Intelligent Transportation Systems Magazine, 14*(2), 90–104.

Kazi, M. H., Newaz, S., & Ahsan, M. S. (2018). Design and development of an aircraft type portable drone for surveillance and disaster management. *International Journal of Intelligent Unmanned Systems*, 1–16. https://doi.org/10.1108/IJIUS-02-2018-0004

Khan, M. A., Ectors, W., Bellemans, T., Janssens, D., & Wets, G. (2018a). Unmanned aerial vehicle-based traffic analysis: A case study for shockwave identification and flow parameters estimation at signalized intersections. *Remote Sensing, 10*, 458. https://doi.org/10.3390/rs10030458

Khan, A. S., Ectors, W., Bellemans, T., Janssens, D., & Wets, G. (2018b). Unmanned aerial vehicles for traffic management: A comprehensive overview. *IEEE Transactions on Intelligent Transportation Systems, 19*(4), 1142–1158.

Kim, N. V., & Chervonenkis, M. A. (2015). Situation control of unmanned aerial vehicles for road traffic monitoring. *Modern Applied Science, 9*(5), 1.

Kumar, A., Sharma, K., Singh, H., Naugriya, S. G., Gill, S. S., & Buyya, R. (2021a). A drone-based networked system and methods for combating coronavirus disease (COVID-19) pandemic. *Future Generation Computer Systems, 115*, 1–19.

Kumar, V., Azad, S., Srivastava, S. K., & Giri, D. (2021b). Role of drones in combating the COVID-19 pandemic: A comprehensive review. *Journal of Healthcare Engineering*, Article ID 8829789.

Liew, C. F., DeLatte, D., Takeishi, N., & Yairi, T. (2017). *Recent developments in aerial robotics: A survey and prototypes overview*. Retrieved from https://arxiv.org/pdf/1711.10085.pdf?utm_content=buffer9a5b8&utm_medium=social&utm_source=twitter.com&utm_campaign=buffer

Mbowa, K., Aigbavboa, C., Akinshipe, O., & Thwala, D. W. (2021). An overview of key emerging technologies transforming public transportation in the fourth industrial revolution era. In *International conference on engineering for sustainable world (ICESW 2020)* (IOP Conference series: Material Science and Engineering) (Vol. 1107, p. 012169).

Mehmood, S., Ahmed, S., Kristensen, A. S., & Ahsan, D. (2018). Multi Criteria Decision Analysis (MCDA) of unmanned aerial vehicles (UAVs) as a part of standard response to emergencies. In *4th international conference on green computing and engineering technologies*.

Miller, E. J., & Shaw, S. L. (2001a). Highway traffic congestion and the price of gasoline. *Transportation Research Part D: Transport and Environment, 6*(2), 89–106.

Miller, H., & Shaw, S. L. (2001b). *Geographic information systems for transportation (GIS-T): Principles and applications*.

Mohammed, F., Idries, A., Mohamed, N., Al-Jaroodi, J., & Jawhar, J. (2014a). UAVs for smart cities: Opportunities and challenges. In *2014 international conference on unmanned aircraft systems (ICUAS)*.

Mohammed, M. A., Mohamed, N., Al-Jaroodi, J., & Jawher, A. (2014b). Smart cities: A survey on security concerns. In *2014 international conference on cyber security of smart cities, industrial control system and communications* (pp. 1–6). IEEE.

Olba Albu, A.M. (2006). GIS applications in transportation analysis and planning. A final Term Paper on Introduction to GIS.

Ostojić, G., Stankovski, S., Tejić, B., Đukić, N., & Tegeltija, S. (2015). Design, control and application of quadcopter. *International Journal of Industrial Engineering and Management, 6*(1), 43–48. Retrieved from https://www.researchgate.net/publication/283110601

Outay, F., Mengash, H. A., & Adnan, M. (2020). Applications of Unmanned Aerial Vehicle (UAV) in road safety, traffic and highway infrastructure management: Recent advances and challenges. *Transportation Research Part A: Policy Practice, 141*, 116–129. https://doi.org/10.1016/j.tra.2020.09.018

Patil, P. (2022). Application of deep learning in traffic management: A review. *International Journal of Business Intelligence and Big Data Analytics*, 16.

Photomodeler Technologies. (2020). *Accident scene mapping with a drone*. https://www.photomodeler.com/pm-applications/pub-safety-forensics/drone-ar-mapping/. Accessed Jan 2023.

Pitney Bowes. (n.d.). *MapInfo Pro*. Retrieved from https://www.pitneybowes.com/us/location-intelligence/geographic-information-systems/mapinfo-pro.html

Roberts, M. (2019). Toy drones: A comprehensive overview of features and applications. *Journal of Toy Technology, 14*(3), 67–82.

Salvo, G., Caruso, L., & Scordo, A. (2014). Gap acceptance analysis in an urban intersection through a video acquired by an UAV. *Recent Advances in Civil Engineering and Mechanics*, 199–205. http://www.wseas.us/e-library/conferences/2014/Florence/SEMOTEC/SEMOTEC-26.pdf. Accessed Dec 2021

Sanson, J. S. (2019). Drone use in the construction industry leads to integration into the current civil and construction engineering technology curriculum. In *Proceedings of the 2019 conference for industry and education collaboration*. Retrieved from: https://peer.asee.org/drone-use-in-the-construction-industry-leads-to-integration-into-the-current-civil-and-construction-engineering-technology-curriculum.pdf

Sharma, V., Chen, H.-C., & Kumar, R. (2017). Driver behaviour detection and vehicle rating using multi-UAV coordinated vehicular networks. *Journal of Computer and System Sciences, 86*, 3–32.

Shimonti, A. (2018a). GIS for highways and transport management. In *Geospatial Technologies in Urban System* (pp. 93–113). Springer.

Shimonti, F. (2018b). GIS-based assessment of road traffic accidents in Dhaka city. *Journal of Traffic and Transportation Engineering (English Edition), 5*(1), 67–79.

Shimonti, P. (2018c). *GIS in transportation*. Retrieved December, 2022 on https://www.geospatialworld.net/blogs/gis-in-transportation/

Škorput, P., Mandžuka, S., & Vojvodić, H. (2016). The use of unmanned aerial vehicles for forest monitoring. In *Proceedings of 58th international symposium ELMAR -93–96*.

Smith, J. (2020). Technological advancements in large drones for military applications. *Journal of Military Technology, 25*(3), 45–56.

Smith, R. J. (2021). Evolution and military applications of drones. *Journal of Military Technology, 25*(2), 73–89.

Smith, J., Johnson, A., & Thompson, R. (2020). Comparative analysis of fixed-wing and multi-rotor drones for aerial mapping. *Journal of Unmanned Aerial Systems, 8*(3), 112–125.

Sutton, J. C., Cevllos, F., Faria, D., Kamler, B., Millan, L., Palmerlee, T., Sanchez, T. W., Shiffer, M., Watanabe, W., & Wiggins, W. (2004). *Geographic information systems applications in transit* (p. 55). Transportation Research Board/TCRP Synthesis.

Tatale, O., Anekar, N., Phatak, S., & Sarkale, S. (2018). Quadcopter: Design, construction and testing. *International Journal for Research in Engineering Application & Management, 04*(Special Issue AMET-2018), 1–7. https://doi.org/10.18231/2454-9150.2018.1482

Tatum, M. C., & Liu, J. (2017). Unmanned aircraft system applications in construction. In *Creative construction conference 2017, CCC 2017, 19–22 June 2017, Primosten, Croatia Procedia Engineering* (Vol. 196, pp. 167–175).

United Nations. (2015a). *Concept of sustainable transport: Planning and designing for sustainable and inclusive transportation system*. National capacity building workshop on sustainable and inclusive transportation system held on 9–10 July, Colombo, Sri Lanka.

United Nations. (2015b). *Transforming our world: The 2030 agenda for sustainable development*. United Nations.

Upadhyay, A. (2018a). *Traffic congestion modelling-GIS*. Retrieved January, 2023 from https://www.igismap.com/traffic-congestion-gis-modelling/

Upadhyay, P. (2018b). Application of GIS and remote sensing in accident prone zone identification and accident analysis. *International Journal of Scientific Research in Science and Technology, 4*(2), 679–684.

Zhang, L., Peng, Z., Sun, D., & Liu, X. (2013). A UAV-based automatic traffic incident detection system for low volume roads. In *Presented in 92nd annual meeting*. Transportation Research Board. Paper No. 13–4724.

Zitzman, L. (2018). *Drones in construction: How they're Transforming the Industry*. Retrieved December 5, 2022, from https://www.bigrentz.com/blog/drones-construction

Zulkipli, M. A., & Tahar, K. N. (2018). Multirotor UAV-based photogrammetric mapping for road design. *International Journal of Optics, 2018*. https://doi.org/10.1155/2018/1871058

Chapter 7
Internet of Things (IoT) and Big Data as a Revolutionary Approach for Traffic Management

Oluwayemi-Oniya Aderibigbe and Trynos Gumbo

7.1 General Overview of Internet of Things (IoT)

According to Muthuramalingam et al. (2019), cities worldwide are currently grappling with complex challenges that hinder the achievement of smart city goals, such as outdated transportation planning, environmental pollution, financial management, and security issues. As urbanization and globalization exert pressure on modern cities, there is a growing need to enhance the quality of life for citizens. The development of smart cities necessitates a robust infrastructure and the adoption of cutting-edge technologies. In this context, the emergence of big data and the evolution of Internet of Things (IoT) technologies have played a crucial role in enabling smart city initiatives.

Big data offers cities the opportunity to extract valuable insights from vast amounts of data collected from diverse sources. On the other hand, the IoT facilitates the integration of sensors, radio-frequency identification, and Bluetooth devices into the physical environment, creating highly networked services that provide real-time information. The combination of these technologies, IoT and big data, is pivotal for the development of smart cities and presents new and intriguing challenges for realizing the vision, principles, and requirements of smart city applications. These challenges primarily revolve around technological aspects, aiming to enable cities to embody the main characteristics of a smart environment. The IoT plays a vital role in collecting relevant data, which is then transferred to the cloud via wireless technologies through the internet. This allows users to comprehend various parameters and their impacts, thereby assisting planners in making informed decisions that benefit the overall system operation. Figure 7.1 illustrates the schematic view of IoT technology in transportation and its functioning.

O.-O. Aderibigbe et al., *Emerging Technologies for Smart Cities*,
https://doi.org/10.1007/978-3-031-66943-9_7

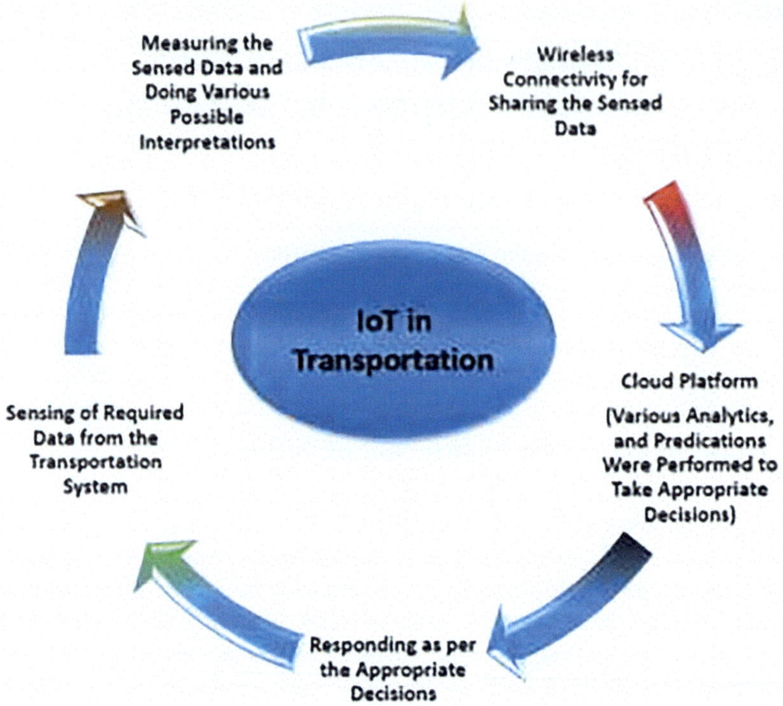

Fig. 7.1 IoT functioning process. (Dash & Kumar, 2017)

7.1.1 Internet of Things (IoT) as a Means of Achieving Sustainable Transport

In recent years, there has been a growing adoption of Internet of Things (IoT) technology in transportation planning, driven by its numerous benefits such as operational efficiencies, cost savings, improved safety, security, and enhanced mobility. The IoT plays a significant role in the development of sustainable transport systems. According to He (2021), the IoT encompasses the concept of connecting all things, where various devices are interconnected to establish seamless communication between people, machines, and objects. It enables the collection and integration of diverse information from sensors and devices, facilitating real-time interconnections. The fundamental characteristic of the IoT lies in the comprehensive perception, reliable transmission, and intelligent processing of data. It encompasses functions such as information acquisition, transmission, processing, and effecting, while also ensuring adaptability, intelligence, privacy, and safety. The integration of wireless technology into transportation systems is expected to have a profound impact on cities and municipalities worldwide. As depicted in Fig. 7.2, this technology enables the implementation of wireless communication networks for traffic

Fig. 7.2 A typical view of IoT Technology for monitoring the activities in the environment. (Source: Birenzvieg 2023)

management, emergency response, and the safety of pedestrians and cyclists (Locke, 2022).

IoT technologies have revolutionized transportation systems by enhancing capacity, improving travel experiences, and ensuring safety, efficiency, and security. Through the use of sensor networks and smart traffic management, local police, emergency services, and government agencies can gain comprehensive visibility across the city, enabling them to effectively manage congestion and respond rapidly to incidents. The integration of IoT-based intelligent transportation systems aligns with the vision of Smart Cities, which leverage advanced communication technologies for efficient city administration and citizen well-being.

Technological advancements have been instrumental in driving the adoption of IoT in transportation. These advancements have led to the development of smarter cities, where the management of fixed and mobile applications has become more streamlined. In fixed applications such as traffic lights, cameras, and intersection management, cellular networks have become more reliable and cost-effective compared to traditional wired networks, presenting new opportunities. Additionally, advancements in mobile router technology and the growth of 4G and 5G networks have significantly improved the capabilities of applications involving police, fire, ambulance, bus, light rail, and paratransit vehicles. These developments have expanded the scope of IoT solutions for transportation, catering to a wide range of needs in diverse operating conditions. As seen in Fig. 7.3, the various applications and descriptions of the IoT in everyday life show its significance, not only in the transportation sector but in all relevant sectors in the economy.

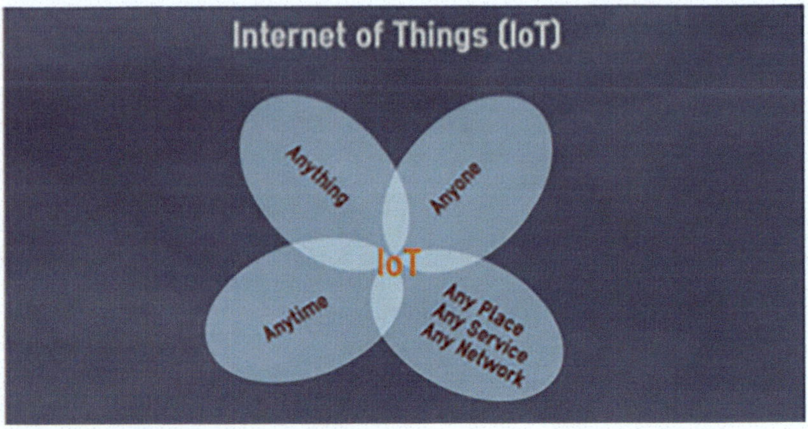

Fig. 7.3 Description of IoT. (Attaran, 2017)

Transportation IoT has paved the way for various applications that significantly improve efficiency, safety, and user experience in the transportation sector. Some key uses of IoT in transportation include:

- Traffic management: IoT applications enable urban traffic management systems to enhance safety and optimize traffic flow, maximizing the value of infrastructure investments (Akhtaruzzaman et al., 2020).
- Public transportation: IoT integration in public transit systems allows transit agencies to operate more efficiently and enhance passenger experience through amenities like informational signage and high-speed Internet connectivity (Lugmayr & Maier, 2021).
- Electric vehicles (EVs) and EV charging: With the rapid growth of EVs and EV charging stations, IoT connectivity plays a crucial role in maintaining the EV infrastructure, handling payment processing, and managing system maintenance (Maresca et al., 2021).
- Railways: IoT solutions are employed in both light rail and heavy commercial rail systems, providing reliable and secure high-speed communication and geo-positioning capabilities, even in challenging environments such as tunnels and urban canyons (Bashir et al., 2021).
- Trucking/logistics: Fleet managers utilize IoT to track vehicle analytics, optimize operations, reduce truck rolls, and automate processes, resulting in cost savings. IoT also facilitates the monitoring and reporting of truck refrigeration systems (Bazaraa & Geissbauer, 2017).

7.1.2 *Internet of Things (IoT) and Smart cities Development*

The evolution of Internet of Things (IoT) technologies has greatly influenced the feasibility and implementation of smart city initiatives. By integrating sensors, radio-frequency identification, and Bluetooth, the IoT enables the creation of highly networked services that interact with the real-world environment. Intelligent Transportation Systems (ITS) are a result of these advancements, offering innovative services for various modes of transport and traffic management, empowering users to make informed decisions and utilize transport networks more efficiently and safely (Muthuramalingam et al., 2019). Several advanced technologies are utilized in Intelligent Transport System planning. These include car navigation systems, traffic signal control systems, container management systems, variable message signs, automatic number plate recognition, and speed cameras. Stakeholders in the transport sector have deployed these technologies to integrate real-time data and feedback from multiple sources, such as parking guidance systems, weather intelligence, and other traffic information, to enhance overall transport operations (Zhou et al., 2019).

Smart transportation encompasses a wide range of technologies, from fundamental management systems like car navigation and traffic signal control to more sophisticated applications for monitoring and controlling traffic. The IoT facilitates better utilization of transportation networks and paves the way for the development of intelligent infrastructure to meet future demands (Khajenasiri et al., 2017). As intelligent transportation systems continue to evolve, they offer a growing array of technological solutions for transportation planners and stakeholders in the built environment, enabling more efficient operation and maintenance of transportation systems. The Intelligent Transportation Society of America highlights the transformative capabilities of Intelligent Transportation Systems (ITS) technology, enabling a wide range of benefits and functionalities:

- Real-time route optimization: ITS technology allows drivers to utilize navigation systems that provide the best routes based on real-time conditions, optimizing travel efficiency (ITS America, n.d.).
- Hazard detection and avoidance: Drivers can receive timely alerts about potentially dangerous situations, allowing them to take preventive measures and avoid crashes (ITS America, n.d.).
- Smart parking guidance: Smart signs and systems assist drivers in locating available parking spaces, streamlining the parking process and reducing congestion (ITS America, n.d.).
- Traffic signal prioritization: Public transit vehicles equipped with ITS technology can communicate with traffic lights to optimize signal timing, reducing delays and improving the efficiency of bus services (ITS America, n.d.).
- Incident detection and response: ITS technology enables the timely detection of traffic incidents, facilitating prompt response and mitigation to ensure safety and minimize disruptions (ITS America, n.d.).

- Dynamic traffic management: Traffic can be rerouted in real-time based on road conditions or emergencies, enhancing overall traffic flow and reducing congestion (ITS America, n.d.).
- Real-time information provision: Travelers can access up-to-date traffic and weather reports, empowering them to make informed decisions and plan their journeys effectively (ITS America, n.d.).
- Fuel consumption management: ITS technology allows drivers to monitor and manage their fuel consumption, promoting fuel efficiency and environmental sustainability (ITS America, n.d.).
- Adaptive speed limits and signal timing: ITS systems can adjust speed limits and signal timing based on real-world conditions, improving safety and optimizing traffic flow (ITS America, n.d.).
- Enhanced freight management: ITS technology contributes to improved tracking, inspection, safety, and efficiency in freight transportation, enhancing the overall supply chain (ITS America, n.d.).
- Improved public transportation: ITS advancements make public transportation more convenient and reliable, enhancing the overall transit experience for passengers (ITS America, n.d.).
- Infrastructure monitoring: ITS systems can monitor the structural integrity of infrastructure, such as bridges, facilitating proactive maintenance and ensuring safety (ITS America, n.d.).

7.1.3 Types of IoT Technologies in Intelligent Transportation Systems

There are various IoT technologies utilized in intelligent transportation systems, each serving specific purposes:

I. Positioning Technology: GPS and Wi-Fi are examples of positioning technologies that provide location and traffic information for road users. These technologies have wide-ranging applications for both indoor and outdoor activities, aiding in positioning requirements and facilitating intelligent transportation (Ding et al., 2020).

II. Tracking Technology: Vehicle monitoring involves using video image processing methods to track vehicles in terms of traffic flow, speed, lane change, and trajectory. This real-time monitoring helps understand traffic conditions and driver behavior, contributing to enhanced safety (Atzori et al., 2010). Monitoring systems employing SoftSIM technology have been developed to improve public traffic safety through real-time tracking and authenticated communication (Yadav & Rishi, 2021).

III. Network Communication Technology: Vehicle-to-vehicle (V2V) and vehicle-to-infrastructure (V2I) communication technologies, such as 5G, play a crucial role in addressing various traffic problems. They facilitate optimization models for street load balancing and enable vehicle positioning, location node acquisition, and roadside sensor intersection nodes (He, 2021).

IV. Vehicle and Road Condition Management Technology: This includes vehicle formation control and real-time traffic prediction. The Internet of Things platform and wireless sensor networks are utilized to enable vehicle coordination and real-time environmental information collection. Deep learning algorithms are employed to address multi-robot collaborative navigation and control (Guan et al., 2020).

V. Vehicle and Traffic Prediction: Real-time road traffic forecasting relies on large amounts of data for accurate analysis. Multivariate spatial-temporal autoregressive (MSTAR) models are used to capture the transient behavior of the traffic network and generate continuous, real-time outputs (Min & Wynter, 2011).

7.1.4 Role of Internet of Things (IoT) Technologies in Transportation Management

The role of Internet of Things (IoT) technologies in transportation management is significant, providing solutions to various transportation challenges. IoT enables enhanced communication and information distribution, facilitating integration in personal, public, and commercial vehicles (Smith, 2022). It extends its capabilities to different elements of transportation, such as optimizing engine performance, fuel efficiency, destination and route planning, parking management, and more. A comprehensive understanding of the applications and significance of IoT in transportation can be found in the article referenced below.

7.1.4.1 IoT and Fleet Management

A distributed IT system is employed to effectively oversee and manage public transportation fleets, encompassing various components and infrastructure elements (see Fig. 7.4). This system encompasses the coordination of vehicles, the control center, depot management, and key features of the road/track infrastructure, including traffic signals and switches, all essential for efficient operations management. Furthermore, fleet management extends beyond these aspects to include passenger guidance, information, and communication systems such as information displays, public announcement speakers, passenger counting devices, as well as gates and turnstiles, enhancing the overall passenger experience and ensuring seamless operations (Tran-Dang et al., 2020).

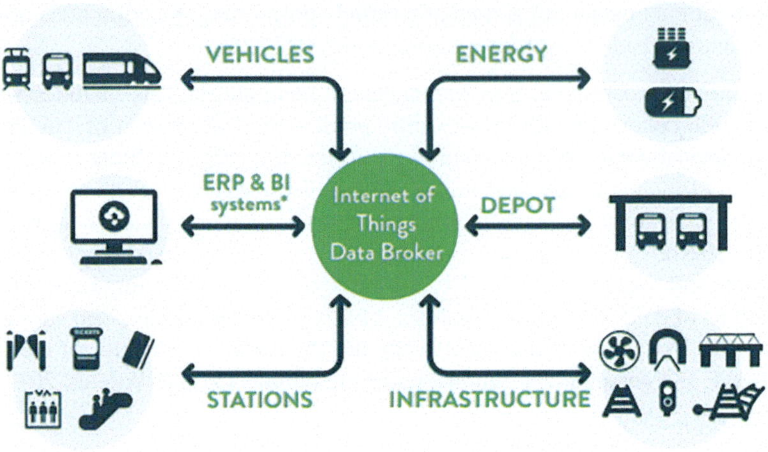

Fig. 7.4 Internet of Things and Fleet Management. (Ushakov et al., 2022)

7.1.4.2 IoT and Transport Infrastructure Management

The effective management of transportation infrastructure is enhanced through the utilization of the Internet of Things (IoT), which enables the remote monitoring and control of traffic light situations using data derived from connected vehicles (Chen et al., 2020). A key application of IoT in this context is the utilization of traffic volume data obtained from dedicated short-range communications systems (DSRC) to regulate traffic light operations (Yang & Qian, 2019). By installing networks of DSRC devices along roadways, traffic congestion can be monitored, leading to informed adjustments in light patterns at major intersections (Deloach et al., 2018). Additionally, IoT technology provides valuable information to drivers of connected vehicles, including scheduled refueling times, which aids in preventing unnecessary delays (Chen et al., 2020). Furthermore, IoT facilitates the provision of real-time information to passengers regarding bus locations, departure times, delays, and other pertinent last-minute updates, enhancing the overall transportation experience (Yang & Qian, 2019).

7.1.4.3 Safety

Safety is a crucial aspect of transportation that can be enhanced through the implementation of IoT technologies. One application of IoT in this domain is the design of steering wheel control systems, which provide drivers with essential information regarding their speed, distance traveled, and the need for rest breaks (Blanco, 2023). These systems can also issue warnings to drivers or their managers if speed limits are exceeded. Furthermore, IoT enables the integration of road and environmental data, alerting drivers to potential hazards such as sharp curves, landslides, or animals on the road. The primary objective of these IoT-enabled safety measures is to proactively notify drivers of impending dangers, thereby preventing accidents.

7.1.4.4 IoT and Traffic Management

The data collected by sensors on traffic status and volumes play a significant role in regulating vehicle traffic, leading to improved safety and the efficient design and maintenance of infrastructure (see Fig. 7.5) (Blanco, 2023). These data-driven insights enable traffic management authorities to make informed decisions and implement measures that enhance road safety and optimize transportation systems. Moreover, in the context of corporate sustainability initiatives, IoT devices contribute to reducing pollution and improving the environment. By providing real-time information on traffic situations, drivers and road users can make informed decisions about the appropriate timing for their trips, thereby reducing the number of vehicles on the road and minimizing the emission of carbon dioxide into the environment (Blanco, 2023) (Fig. 7.6).

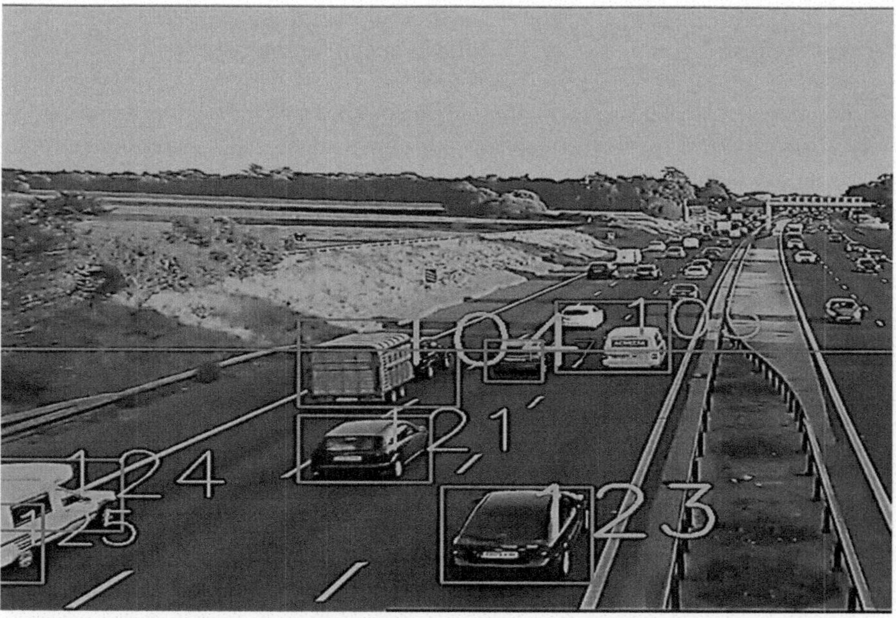

Fig. 7.5 Vehicle detection using IOT technology. (Source: Javaid et al., 2018)

Fig. 7.6 A typical traffic situation on the road

7.1.4.5 IoT and Management of Public Transportation

IoT has played a pivotal role in revolutionizing public transportation systems worldwide, primarily by providing valuable insights into user statistics and enabling route optimization. By utilizing IoT sensors integrated into vehicles and security cameras, data can be collected to optimize existing routes or design new ones. The information derived from these data sets can be used to develop long-term strategies and provide real-time alerts, such as notifying users about vehicle occupancy. This information empowers passengers to plan their journeys more effectively, encouraging the use of large public transport vehicles and discouraging single occupancy vehicles that contribute to pollution and exacerbate global warming by accommodating fewer individuals.

7.1.4.6 IoT and Automobiles Efficiency

The IoT industry has revolutionized the capabilities of conventional cars, particularly through the integration of IoT technology into "smart cars." These intelligent vehicles not only offer the standard functionalities of regular cars but also enhance their features. Smart cars are equipped with sensors that enable them to detect changes in the surrounding environment, providing real-time information about road conditions and traffic congestion. Additionally, they incorporate thermostats that automatically adjust the internal temperature for optimal comfort. IoT devices have also facilitated the implementation of child lock features, ensuring that car doors remain locked while in motion, preventing any accidental unlocking (See Fig. 7.7).

7.1.4.7 IoT and Rail Management

In the realm of rail transportation, many developing countries still rely on traditional technology systems that often suffer from sluggish performance. However, the implementation of IoT systems has proven to deliver improved performance, monitoring, and control, facilitating enhanced execution, issue handling, and improvisation (Techvidvan, 2021). Furthermore, in the case of Mass Rapid Transit (MRT) systems, IoT integration helps elevate them from being mere transportation options to becoming comprehensive and efficient services. IoT enables the implementation of better standards, increased output, and improved services such as scheduling timings, customer care, quick response, management, and cleanliness.

7.1.4.8 Inventory Management System

The advent of smart inventory management systems has revolutionized the way inventory is monitored and managed by providing real-time information about inventory conditions remotely. This advancement is particularly valuable in supply chain management, as it enables the generation of vast amounts of data and establishes an integrated framework for efficient inventory management (Techvidvan, 2021). Consequently, IoT plays a pivotal role in optimizing logistics, inventory management, and supply chains (Fig. 7.8).

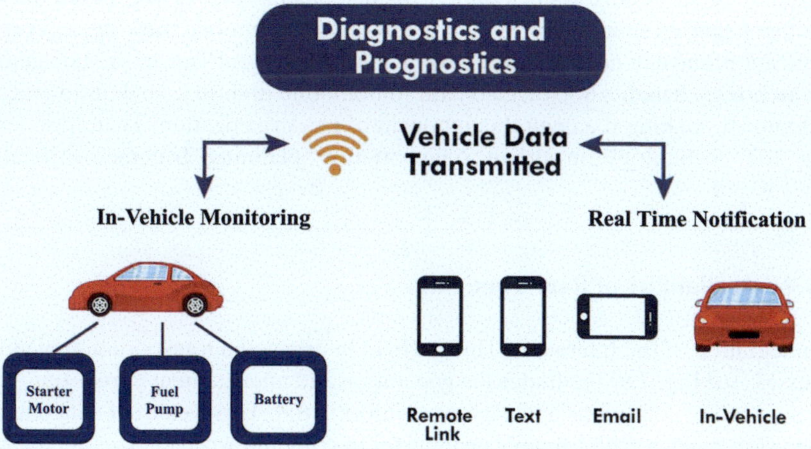

Fig. 7.7 IoT sensors and operating devices in a typical Automobile

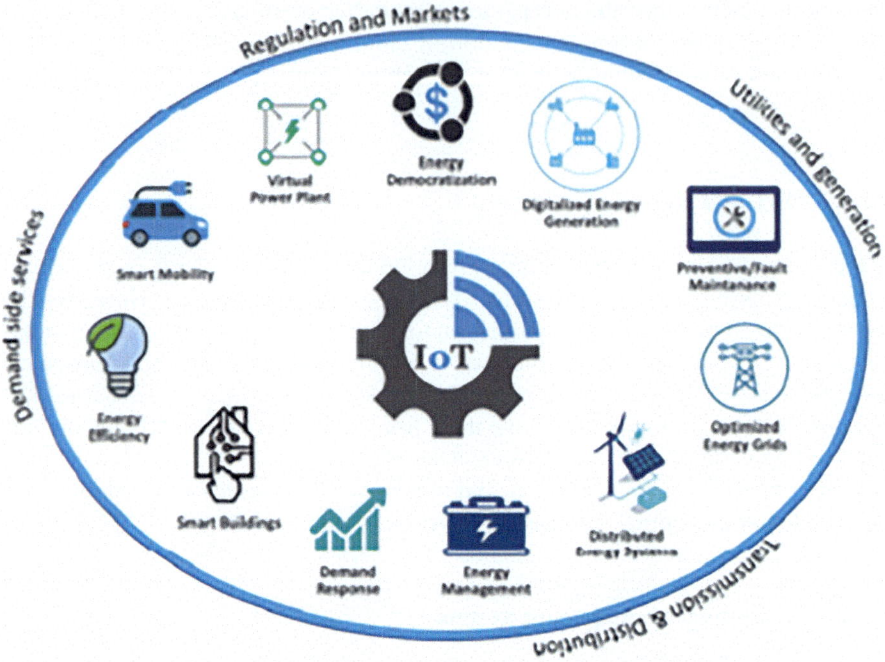

Fig. 7.8 An integrated smart energy system using IoT. (Motlagh et al., 2020)

Transportation serves as a crucial link between consumers and goods, facilitating their movement from origin to destination and playing a significant role in individual and societal development. Recognizing the importance of transportation, it becomes essential to invest in technology to enhance efficiency and effectiveness. The Internet of Things (IoT) emerges as a valuable tool in this regard, enabling the integration of the entire transportation management process. IoT offers various advantages such as smart mobility, smart buildings, automobile tracking, safer route optimization, and extended distance coverage. An Intelligent Transport Management System brings benefits to travelers, the automobile industry, and infrastructure development, making it a central component of any organization. Therefore, leveraging IoT technology in urban transportation planning becomes critical in today's world.

7.1.4.9 Availability of Real-Time Data

The integration of the Internet of Things (IoT) in urban and transportation planning processes (see Fig. 7.9) facilitates the generation and management of real-time data, including traffic conditions and passenger information. This wealth of data empowers decision-makers with valuable insights for making informed choices and enables

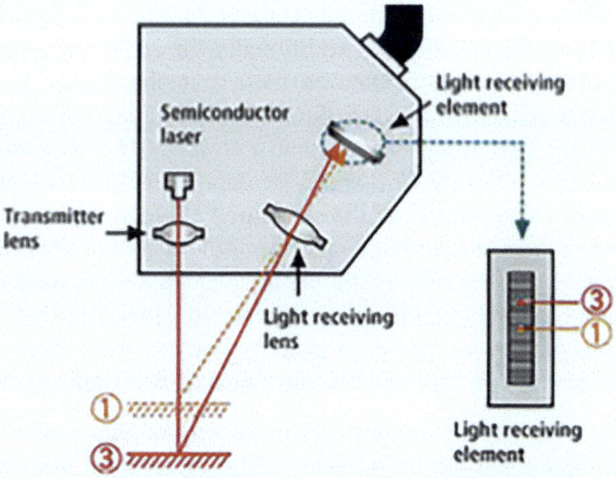

Fig. 7.9 A laser displacement sensor. (Gopinath & Ramadoss, 2021)

efficient monitoring of transportation performance, ultimately leading to more effective and streamlined transport delivery.

7.2 The Role of Big Data in Transportation

The utilization of big data presents cities with a valuable opportunity to gain insights from a vast amount of information collected from diverse sources. This exponential growth and rapid flow of exploitable open data will bring about significant and transformative changes across multiple sectors, with transportation being a prominent one. The International Transport Forum (ITF, 2015) emphasizes the compelling importance of big data analytics in urban planning, particularly in the convergence of high-definition geographic data with information on citizens' observed or interpreted use of urban space and intelligent transport. Real-time analysis and visualization of transport network usage enable improved safety planning by processing data on vehicle operations and the surrounding environment, thereby avoiding or minimizing potential conflicts. Big data represents the amalgamation of novel data collection mechanisms facilitated by ubiquitous digital devices, expanded storage capacity, increased computing power, and advanced sensing and communication technologies. Some of these data streams are specifically designed to address everyday challenges and tasks. For example, data obtained from automatic toll payment transponders is vital for efficient toll processing and secure payment.

According to ITF (2015), Big Data refers to the collection, storage, and interpretation of extremely large datasets with the help of advanced technology. While there is no universally agreed-upon definition, it is commonly understood as datasets that are too large to be managed or analyzed using the resources of a typical personal computer or commonly used spreadsheet applications. Riahi and Riahi (2018) define Big Data as the evolution and utilization of technologies that deliver the right information to the right user at the right time from a massive, exponentially growing dataset in our society. Additionally, in 2011, the International Data Corporation report defined Big Data as modern technologies and architectures designed to extract value from vast volumes of diverse data through high-velocity capture, discovery, and analysis (Alexandru et al., 2016). Key attributes of Big Data include volume, velocity (the speed of data collection and processing), and variety (the range of structured and unstructured elements in the dataset). Although these attributes differentiate Big Data from other types of data, it is crucial to recognize that they do not capture the fundamental changes that have led to the emergence of such large and exploitable datasets in recent years. Big Data is not a standalone concept; rather, it encompasses a process that involves data acquisition, processing, and interpretation. The specific attributes of Big Data are elaborated below:

- Volume: This pertains to the sheer amount of data being generated, stored, and processed within a system. The increase in volume is driven by the growing amount of data being generated and stored, as well as the need to effectively utilize it.
- Variety: This encompasses the diverse range of data types managed by an information system. The proliferation of data types leads to intricate connections and relationships between these data, adding to the complexity. The variety aspect also relates to the potential applications associated with raw data.
- Velocity: This refers to the speed at which data is generated, captured, and shared. Data arrives in a continuous stream and is typically analyzed in real-time to enable prompt decision-making.
- Veracity: This indicates the reliability and credibility of the data. The quality of big data is challenging to control due to its diverse sources, which may lack guarantees regarding content accuracy and presentation format. Experienced data analysts are required to assess data compliance, accuracy, and truthfulness. This evaluation should consider the responsibilities of the initial data generator, the purpose of the data, and the reactions of the data recipients.
- Variability: This relates to the inherent variability in the meaning of data, particularly evident in natural language processing. Companies must develop sophisticated programs capable of understanding context and decoding the precise meaning of words to effectively handle such variability.
- Visualization: This pertains to the readability and accessibility of data. Achieving comprehension and actionability requires the utilization of numerous spatial and temporal parameters and their relationships, enabling data to be presented in a manner that is easily understandable and actionable.
- Value: This signifies the capacity of data to provide new insights and contribute to knowledge creation, unlocking its potential to deliver meaningful benefits.

7.2.1 *Big Data Analytics*

The emergence of Big Data as a valuable asset highlights the need for appropriate tools and methodologies to analyze and extract meaningful patterns from extensive datasets. The increasing diversity and speed of data processing pose new challenges for analyzing structured data. Mere data analysis and report generation are no longer sufficient; the existing systems must facilitate data analysis effectively (Riahi & Riahi, 2018).

Big Data Analytics encompasses the collection, organization, and analysis of large datasets to unveil various patterns and derive valuable insights. It encompasses a range of technologies and techniques that require innovative integration approaches to unlock hidden value from complex and vast datasets. The primary goal of Big Data Analytics is to address both current and emerging issues in a more efficient and effective manner.

7.2.2 *Types of Big Data Analytics*

(a) Descriptive Analytics: This initial stage of data processing involves examining historical data and posing questions to uncover patterns and gain insights. It also utilizes data mining techniques to organize data and reveal trends and probabilities that inform future occurrences.
(b) Diagnostic Analytics: This type of analytics aims to identify the root causes of events and behaviors, providing answers to the fundamental questions underlying a problem.
(c) Predictive Analytics: Focused on forecasting, predictive analytics leverages past data to predict future events. It employs techniques such as data mining and artificial intelligence to analyze current data and generate scenarios of potential future outcomes.
(d) Prescriptive Analytics: Prescriptive analytics goes beyond prediction and provides actionable solutions to problems. It addresses the question of what should be done, utilizing historical and predictive analytics parameters to determine the best course of action.

Overall, big data systems serve as valuable sources of information for public organizations and citizens. These systems gather data from various sources, including paper and digital documents, government networks, websites, social media, and operational systems. The information provided by big data systems excludes personal or confidential data protected by control mechanisms.
According to Gartner Inc. (**2012**), big data can be classified into five categories:

1. Operational data: Customer, supplier, partner, and employee information from transactions and databases;

2. Dark data: Unstructured data accumulated over time, such as emails and multimedia;
3. Commercial data: Data obtained from aggregators (which reads RSSs);
4. Public data: Data from public institutions (the Government and the Ministries); and
5. Data from social media: Activity on blogs and social networks, useful for identifying trends, attitudes, and preferences. Analysing big data can help identify business trends, prevent diseases, and combat crime.

7.2.3 Challenges of Big Data

Managing huge volumes of data poses significant challenges in the big data era. Wang et al. (2018) highlight several major challenges associated with data management:

- Heterogeneity of Data: Data in big data environments come from diverse sources, making them heterogeneous. Before performing analyses, it is essential to structure the data to ensure compatibility and consistency across different sources.
- Data Completeness: Incomplete data can create problems during analysis. To mitigate this issue, missing values in numerical data need to be appropriately handled by inserting null values in place of missing values to avoid affecting the integrity of the data.
- Processing Time: The sheer volume of data makes processing time a critical challenge. As data volumes increase, the time required for processing and analysis also increases. Efficient strategies and technologies are required to handle and process data within reasonable time frames.
- Real-Time Processing: The high velocity of data arrival presents a challenge for real-time processing. The ability to process and analyze incoming data at a faster rate than it arrives is crucial to extract timely insights and make informed decisions.
- Scalable and Fault-Tolerant Storage: As data grows exponentially, scalable storage solutions are necessary to accommodate increasing volumes. The storage infrastructure should also be fault-tolerant to ensure data integrity and availability.
- Data Privacy and Security: With the increasing volume of data, ensuring data privacy becomes paramount. Strict access control mechanisms should be implemented at different stages of the big data lifecycle to protect sensitive information. Data sharing should be restricted to ensure that only authorized users have access to the necessary data for business knowledge, and proper safeguards should be in place when data is shared for analysis.

7.2.4 Role of Big data in Transportation Planning and Management

The rapid urbanization and population growth have presented various transportation challenges, including traffic congestion and accidents, putting strain on the environment. To address these issues, the implementation of intelligent transportation systems has become crucial. By combining electric sensor technology with information technology, intelligent transportation systems have the potential to greatly improve transportation management. Therefore, it is essential to gather diverse traffic data accurately and in a timely manner and employ big data technology to construct traffic data models, as this serves as a fundamental requirement for achieving intelligent transportation solutions.

7.2.4.1 The Application of Big Data Technology in Traffic Management

The incessant traffic congestion on major roads in urban centers can be attributed to several factors, including high traffic volumes exceeding the road capacity, suboptimal road infrastructure, insufficient parking facilities, inconsistent traffic planning, and the rising number of vehicles. The construction activities in cities have also contributed to the occupation of roads, exacerbating the imbalance between road capacity and traffic demand.

Big data technology offers significant advantages in the collection and monitoring of traffic data. Firstly, it enables more efficient monitoring of the entire traffic network system by providing real-time information on traffic situations (Hou, 2017). Secondly, big data technology focuses on capturing data from various sources, including human interactions, providing a more comprehensive understanding of traffic patterns. Thirdly, it overcomes the limitations of data storage, allowing for improved collection and conversion of simulated data. Finally, big data technology facilitates real-time observations and collection of traffic flow information, enabling informed judgments and assumptions about traffic conditions.

7.2.4.2 The Application of Big Data Technology in Transport Safety

The application of big data technology in ensuring the safety of road users is of immense importance. Urban centers have experienced a continuous increase in traffic demand, leading to various transport challenges, including accidents. Stakeholders in the transportation sector have recognized the significance of road traffic safety and have made it a priority to reduce the occurrence of accidents and save lives. Accidents in the road traffic system can be attributed to multiple factors such as weather conditions, driver quality, and the environment.

The predictive capabilities and performance of big data technology can play a crucial role in actively forewarning accidents, allowing relevant parties to assess the probability of accident occurrence. For instance, the utilization of GPS positioning technology enables the collection of vehicle travel data, providing insights into vehicle conditions (Hou, 2017). Furthermore, the implementation of road environment and facility supervision systems facilitates the timely collection of dynamic information regarding the road environment and infrastructure. This information can be disseminated through traffic radio platforms or delivered to vehicles within the driving region via mobile phone messages, utilizing cloud computing technologies. Additionally, by integrating big data technology with emergency aid systems, accurate information regarding accident locations can be provided, facilitating prompt responses from medical and fire rescue teams.

7.2.4.3 The Application of Big Data Technology in Optimizing the Public Transportation System

Public transportation, comprising rail transit, buses, and cableways operating on designated routes within urban areas, faces challenges due to limited resources, resulting in issues such as prolonged waiting times for passengers and overcrowded public transport. However, the integration of big data technology can significantly enhance the efficiency of resource allocation and promote the rational utilization of public resources. Real-time monitoring of passenger flow volume at stations enables effective resource management (Li et al., 2020). Mobile applications can provide passengers with real-time information regarding bus operations, allowing them to make informed decisions about their travel plans. These applications can provide details on passenger flow volume, allowing passengers to adjust their schedules accordingly (Li et al., 2020). As a result, waiting times at bus stations can be reduced when passengers have access to information on departure times.

7.2.4.4 The Application of Big Data Technology in High-Speed Rail

Public transportation, specifically the rail system, plays a vital role in modern cities. The utilization of big data technology in this context offers several benefits to both passengers and taxi drivers. By leveraging big data, passengers can access essential information about the rail system, including arrival and departure times and passenger load. This information empowers passengers to plan their journeys more effectively and reduces waiting times at train stations (Yang et al., 2020). Through the application of big data technology, taxi drivers can establish collaborations with high-speed transportation departments, enabling them to obtain advance information such as phone numbers and train schedules of passengers. This facilitates accurate prediction of travel times for passengers from their origin to the high-speed rail station. Additionally, big data technology provides drivers with real-time information for optimal route selection, particularly in unpredictable situations like accidents or adverse weather conditions. Consequently, drivers can be guided to safer and faster routes, enhancing overall efficiency (Yang et al., 2020).

Furthermore, the integration of GPS positioning technology in mobile phones enables passengers to conveniently access information regarding nearby taxi services at the desired pick-up location and time. This real-time transportation information service assists taxi drivers in arriving at stations via the most efficient routes in the shortest possible time (Chen, 2016a, b).

7.2.4.5 Big Data and Active Parking System

In recent years, the surge in car ownership has led to an increase in parking difficulties within urban areas. However, the application of big data technology in intelligent transportation systems offers a potential solution to this problem. By utilizing big data technology, transportation service information can be actively disseminated to users, aiding in the management of parking-related issues (Hou, 2017). For example, through the utilization of electronic license plate GPS positioning technology, real-time access to information about the duration and precise locations of parked cars becomes possible. Additionally, the implementation of parking guidance systems can provide drivers with valuable information regarding the nearest available parking lots. This not only alleviates the challenge of searching for suitable parking spaces but also reduces travel time and enhances fuel efficiency. Furthermore, mobile applications can be employed to obtain vehicle location data and access parking-related information conveniently (Hou, 2017).

7.2.5 *Big Data Technologies*

Big data technology encompasses various technologies and system architectures that aim to extract social or economic value from high velocity, large scale, and diverse data streams (Cavoukian & Jones, 2012a, b). Four interconnected technological developments form the core of big data: Firstly, new techniques have emerged to efficiently process large data sets within the limitations of available computing capacity. Map-reducing work processes, such as Hadoop or its derivatives, leverage parallel processing to break down complex semi-structured and unstructured data sets into manageable subsets. These algorithms are scalable and can be directly applied to data regardless of its size, format, or complexity, without the need to formalize database relationships in advance (Cavoukian & Jones, 2012a, b).

In the context of transportation systems, information is gathered from cameras and microcontrollers to optimize public transit, monitor the environment, and enhance security applications (Calabrese et al., 2011a, b). The information for Intelligent Transportation Systems (ITS) is extracted from two types of sensor networks: fixed sensor agents and dynamic probes, such as mobile sensors (Calabrese et al., 2011a, b). Fixed sensors, permanently installed, monitor real-time dynamics of the surrounding terrain and provide information such as vehicle speed, distance between vehicles, road surface conditions, and CO_2 emissions. Fixed sensors can be connected via wired or wireless methods, including roadside traffic cameras and various sensor types like radar, ultrasound, or infrared radiation (Calabrese et al., 2011a, b). On the other hand, mobile sensors, including GPS-enabled vehicles, transmit updated location coordinates along with each transmission (Ratti & Nabian, 2010a, b).

Both fixed sensors and GPS receivers require substantial investment for installation and maintenance. To achieve adequate sensing capability, a large number of sensors are often deployed. The accuracy of traffic estimation using GPS receivers depends on the density of reporting vehicles in each area, and as the number of probes increases, transmission costs to the central system can become significant (Ratti & Nabian, 2010a, b). Fixed sensors, while providing valuable data, also incur high costs due to their wired connection infrastructure, which is necessary to create a comprehensive view of mobility conditions in a city (Ratti & Nabian, 2010a, b).

In summary, big data and the Internet of Things (IoT) are pivotal in sustainable planning and smart city development. The IoT refers to the interconnection of devices and networks using standard internet protocols, enabling communication and data exchange. Big data, on the other hand, pertains to numerical data generated through the use of new technologies for personal or professional purposes. The integration of IoT and big data presents significant opportunities and benefits for the transportation sector and the built environment. Therefore, the adoption of IoT and big data in transportation planning is highly recommended to enhance effectiveness and profitability (Cavoukian & Jones, 2012a, b).

References

Akhtaruzzaman, M., Rahman, M. M., & Anwar, S. (2020). Internet of Things (IoT) applications in transportation: A systematic review. *IEEE Access, 8*, 8692–8705.

Alexandru,A., Alexandru, C.A., Coardos, D. & Tudora, E. (2016). Big data: Concepts, technologies and applications in the public sector International Journal of Computer, Electrical, Automation, Control and Information Engineering Vol:10, No:10, 1732–1738.

Attaran, M. (2017). The internet of things: Limitless opportunities for business and society. *Journal of Strategic Innovation and Sustainability, 12*(1), 10–29.

Atzori, L., Iera, A., & Morabito, G. (2010). The internet of things: A survey. *Computer Networks, 54*(15), 2787–2805.

Bashir, A. K., Bilal, M., Raza, Z., & Zia, T. (2021). Intelligent and scalable IoT platform for the railway system. *IEEE Access, 9*, 22542–22559.

Bazaraa, A., & Geissbauer, R. (2017). Internet of things in logistics. In E. Friess (Ed.), *Internet of things* (pp. 235–248). Springer.

Birenzvieg, A. (2023). *Behaviour Detection*. Assessed on June 2023 from https://www.security-magazine.com/keywords/3014-behavior-detection

Blanco, J.M. (2023). *IoT and transportation: How technology is improving the way we travel*. Accessed in February, 2023 from https://www.plainconcepts.com/iot-transportation/

Calabrese, F., Colonna, M., Lovisolo, P., & Ratti, C. (2011a). Real-time urban monitoring using cell phones: A case study in Rome. *IEEE Transactions on Intelligent Transportation Systems, 12*(1), 141–151.

Calabrese, F., Pereira, F. C., Di Lorenzo, G., Liu, L., & Ratti, C. (2011b). The geography of taste: Analyzing cell-phone mobility and social events. *IEEE Pervasive Computing, 10*(4), 36–44.

Cavoukian, A., & Jones, J. (2012a). *Privacy by design in the age of big data*. Office of the Information and Privacy Commissioner for the Province of Ontario.

Cavoukian, A., & Jones, D. (2012b). *Big data and innovation, setting the record straight: De-identification does work*. Information and Privacy Commissioner of Ontario.

Chen, J. (2016a). The application of big data technology in intelligent transportation. *Goods and Quality, 2016*(42), 404.

Chen, L. (2016b). Design and implementation of a real-time carpooling system based on big data analysis. *Mobile Information Systems, 2016*, 4901347.

Chen, L., Wang, J., & Lv, Y. (2020). Traffic signal control strategy based on internet of things. *Journal of Physics: Conference Series, 1517*(1), 012083.

Dash, A. & Kumar, N.M. (2017). *The internet of things: An opportunity for transportation and logistics*. Proceedings of the international conference on inventive computing and informatics (ICICI 2017) IEEE Xplore compliant - part number: CFP17L34-ART, ISBN: 978-1-5386-4031-9.

Deloach, S. A., Park, J., & Nandiraju, R. S. (2018). Advanced traffic signal control in connected vehicle environments. *Transportation Research Part C: Emerging Technologies, 86*, 142–159.

Ding, J., Peng, H., Zhang, Y., & Li, L. (2020). Penetration effect of connected and automated vehicles on cooperative on-ramp merging. *IET Intelligent Transport Systems, 14*(1), 56–64.

Gartner Inc. (2012). *Big data strategy components: Business essentials*. Accessed on September 2020, Retrieved from: https://www.gartner.com/doc/2191415/bigdata-strategy-components-business

Gopinath, V. K., & Ramadoss, R. (2021). Review of structural health monitoring for restoration of heritage buildings. *Material Today: Proceedings, 43*(2021), 1534–1538.

Guan, J. S., Zhou, W. D., Kan, S. B., Sun, Y., & Liu, Z. B. (2020). Robot formation control based on internet of things technology platform. *IEEE Access, 1,1 PP(99)*.

He, S. (2021). Application of Internet of Things technology in intelligent transportation in post-epidemic era. *Advances in economics, business and management research,* proceedings of the 2021 3rd international conference on economic management and cultural industry (ICEMCI 2021) Vol. 203, pp. 3007–3011.

Hou, Y. (2017). The application discussion of big data technology in intelligent transportation. In *2nd International Conference on Computer Engineering, Information Science & Application Technology (ICCIA 2017)*. Advances in Computer Science Research, volume 74, pp. 440–443.

Intelligent Transportation Society of America. (n.d.). *ITS benefits*. Retrieved from https://www.itsa.org/advocacy/benefits-of-its

International Transport Forum-ITF. (2015). *Big data and transport: Understanding and assessing options*. Accessed on January, 2022, from https://www.itfoecd.org/sites/default/files/docs/15cpb_bigdata_0.pdf

Javaid, S., Sufian, A., Pervaid, S., & Tanveer, M. (2018). Smart traffic management system using Internet of Things. *International Conference on Advanced Communications Technologies (ICACT) 2018*, 393–398. https://doi.org/10.23919/ICACT.2018.8323770

Khajenasiri, I., Estebsari, A., Verhelsta, M., & Gielena, G. (2017). A review on Internet of Things solutions for intelligent energy control in buildings for Smart City Applications. In *8th international conference on sustainability in energy and buildings, SEB-16, 11–13 Sept* (Vol. 111, pp. 770–779).

Li, S., Deng, Z., & Liu, X. (2020). Integration of big data technologies in public transport management: A systematic review. *IEEE Access, 8*, 212365–212378.

Locke, J. (2022). *IoT in transportation: Solutions and applications*. Accessed in December, 2022 from https://www.digi.com/blog/post/iot-solutions-for-transportation

Lugmayr, A., & Maier, R. (2021). Public transport's role in smart cities: A literature review. *Transportation Research Interdisciplinary Perspectives, 10*, 100390.

Maresca, P., Kounelis, I., & Carli, R. (2021). Assessing the effects of urban consolidation centers on electric vehicle charging infrastructure planning. *Sustainability, 13*(6), 3389.

Min, W., & Wynter, L. (2011). Real-time road traffic prediction with spatio-temporal correlations. *Transportation Research Part C: Emerging Technologies, 19*(4), 606–616.

Motlagh, N. H., Mohammadrezaei, M., Hunt, J., & Zakeri, B. (2020). Internet of things (IoT) and the energy sector. *Energies, 13*, 494. https://doi.org/10.3390/en13020494

Muthuramalingam, S., Bharathi, A., Rakesh Kumar, S., Gayathri, N., Sathiyaraj, R., & Balamurugan, B. (2019). IoT based intelligent transportation system (IoT-ITS) for global perspective: A case study. In V. E. Balas et al. (Eds.), *Internet of Things and big data analytics for smart generation, intelligent systems reference library*. https://doi.org/10.1007/978-3-030-04203-5_13

Ratti, C., & Nabian, N., (2010a). Virtual space, the city to come. In: *Innovation perspectives for the 21st century* (pp. 383–397). S.L.:BBVA.

Ratti, C., & Nabian, N. (2010b). Using big data to produce new mobility indicators: Case study in Boston. *Transport Policy, 18*(6), 528–534.

Riahi, Y., & Riahi, S. (2018). Big data and big data analytics: Concepts, types and applications. *International Journal of Research and Engineering., 5:(9)*, 524–528. https://doi.org/10.21276/ijre.2018.5.9.5

Smith, J. (2022). Internet of things in transportation: Applications and role. *International Journal of Transportation Science and Technology, 6*(2), 145–163.

Techvidvan. (2021). *Roles and applications of IoT in transportation*. Accessed in January, 2023 from https://techvidvan.com/tutorials/role-and-applications-of-iot-in-transportation/

Tran-Dang, H., Krommenacker, N., Charpentier, P., & Kim, D. S. (2020). Toward the internet of things for physical internet: Perspectives and challenges. *IEEE Internet of Things Journal, 7*(6), 4711–4736.

Ushakov, D., Dudukalov, E., Kozlova, E. & Shatila, K. (2022). The Internet of Things Impact on Smart Public Transportation. *Transportation Research Procedia, 63*, 2392–2400.

Wang, Y., Kung, L. A., & Byrd, T. A. (2018). Big data analytics: Understanding its capabilities and potential benefits for healthcare organizations. *Technological Forecasting and Social Change, 126*, 3–13.

Yadav, S., & Rishi, R. (2021). Secure and authenticate communication by using soft Sim for intelligent transportation system in smart cities. *Journal of Physics: Conference Series, 1767*(1), 012049.

Yang, D., & Qian, Y. (2019). Internet of things-based traffic light control for emergency vehicles in smart cities. *Sensors, 19*(7), 1562.

Yang, Q., Chen, J., Deng, Z., & Li, S. (2020). Application of big data in the field of transportation: A comprehensive review. *IEEE Access, 8*, 57571–57583.

Zhou, Y., Wang, J., & Yang, H. (2019). Resilience of transportation systems: concepts and comprehensive review. *IEEE Transactions on Intelligent Transportation Systems, 20*(12), 4262–4276.

Part III
Prospects of Emerging Technologies and the Role of Government and Citizen in Smart Cities and Urban Transportation Development

Chapter 8
Benefits and Challenges of Emerging Technologies in Urban Transportation Planning

Oluwayemi-Oniya Aderibigbe

8.1 Introduction

Throughout history, transportation has been driven by technological advancements, leading to significant improvements worldwide. However, the rapid progress and widespread adoption of technology present both challenges and opportunities for transport operators. The U.S. Department of Transportation, Federal Highway Administration (2013a, b) states that various societal and technological trends have implications for transport operations, often fueling the development of intelligent transport systems (ITS). At a fundamental level, the expansion of telecommunication technologies is revolutionizing multiple aspects of our lives. Mobile communications, for example, generate vast amounts of data that can be transmitted almost instantly. This enables travelers to access relevant information on traffic flow, accidents, weather conditions, and provides convenient platforms to monitor and access real-time transport updates in their surroundings. These advancements in mobile communication are supported by the evolution of mobile communication networks, progressing from third generation (3G/GSM) to fourth generation (4G) technologies in 2011 (U.S. Department of Transportation, Federal Highway Administration, 2013a, b).

In today's world, efficient transport management heavily relies on modern technological solutions. Without proper support from these solutions, the growth and development of the sector can be severely hindered. Mobile applications, as evidenced by data, are widely utilized. For instance, in 2017, mobile application users spent a significant amount of time, totaling 900 trillion hours (Morawski et al., 2019). Additionally, the advent of smartphones has granted customers access to platforms for cargo offers, free cargo space, spot marketing, tender management, warehouse space search and offers, company databases, documents, and instant messaging. Furthermore, the global Internet of Things (IoT) market is projected to grow at an annual rate of approximately 16.7%, reaching a value of USD 1.5 trillion

O.-O. Aderibigbe et al., *Emerging Technologies for Smart Cities*,
https://doi.org/10.1007/978-3-031-66943-9_8

in 2020 and potentially reaching USD 3 trillion by 2024 (Sułkowski et al., 2017a, b). This indicates a significant proliferation of technology across all sectors on a global scale.

8.2 Emerging Transport Technologies

In recent times, technological advancements have led to the emergence of diverse new mobility products and services. Rauch (2017a, b) highlights that digital connectivity not only expands the range of available mobility options but also introduces an entirely new layer of mobility structures. Through data exchange between road users, vehicles, and surrounding infrastructure, a self-regulating system has been developed, enabling real-time traffic planning, on-demand availability, and seamless transitions between different modes of transportation. Briggs and Sundaram (2016a, b) conducted a study predicting that the continued growth of new mobility services like ridesharing and ride-on-demand could lead to the removal of over 20 million vehicles annually from urban roads by 2025. Consequently, one notable trend in transportation technology is the increasing popularity of mobility-as-a-service (MaaS) products. MaaS aims to streamline the entire transportation chain for commuters, providing a seamless experience. Using a smartphone-based system that encompasses all modes of transport, road users can easily plan and book their trips. Several MaaS projects have emerged in Europe in recent years, with the city of Helsinki committing to eliminate the need for privately owned cars by implementing MaaS (Franckx, 2015a, b).

Some Emerging Technologies and their Functions include:

8.2.1 3D Printing

3D printing, also known as Additive Manufacturing (AM), which has its roots in production technology and methods (Ford & Despeisse, 2016; Khajavi et al., 2013a, b, c; Verhoef et al., 2018), emerged in the early twentieth century and the first 3D printer was invented in 1984, evolving from automated production processes. Originally, it found application in the manufacturing and automotive industries. However, in recent years, 3D printing has rapidly grown as a transformative technology in the built environment and urban planning. Its application has proven beneficial in various industries such as medical, aerospace, and construction, enabling mass customization and the production of complex shapes that are difficult to achieve through other means. The term "Rapid Prototyping" (RP) was historically used to describe 3D printing technology, which originated in Japan in the 1980s. In 1984, an American inventor introduced stereolithography (SLA), a 3D printing method that utilizes a liquid photopolymer resin contained in a reservoir (Fig. 8.1).

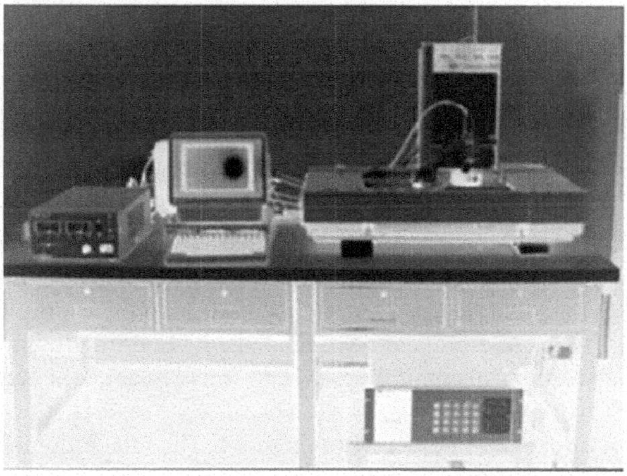

Fig. 8.1 First Invented stereolithography. (Giacomelli & Smedberg, 2014)

The utilization of 3D printing is expected to reshape traffic distribution patterns and reduce transportation volume. Halassi et al. (2019a, b) argue that 3D printing has the potential to disrupt traditional manufacturing and logistics processes as it becomes increasingly available, affordable, and supported by appropriate business models. This increased adoption of 3D printing technology will have implications for transportation and offer new approaches and solutions to existing transportation challenges. Ankner (2014) suggests that 3D printing can facilitate the movement of goods by enabling centralized manufacturing industries to establish small hubs near end users, allowing businesses and individuals to produce their required products on-site. This decentralization of manufacturing could bring 3D printing closer to end users and disrupt traditional manufacturing and transportation planning practices. Transportation planners will need to adapt to changing mobility needs resulting from agile manufacturing strategies. Furthermore, the use of 3D printing can potentially reduce port traffic and long-distance distribution. As more materials become available for 3D printing, there may be a decline in the cargo industry. This shift challenges the long-term strategic plans and financing of ports, which may have to reconsider their facilities and intermodal connections to align with changing demands and adjust congestion management plans.

8.2.1.1 Benefits of 3D printing in the Built Environment

The introduction of 3D printing technology, also known as Additive Manufacturing (AM), has provided solutions to various challenges in the built environment. As a result, stakeholders in the transport sector and the built environment have adopted 3D printing to gain a competitive advantage in the industry.

1. **Sustainable Planning through Waste Reduction**: Sustainable planning aims to promote waste recycling, and 3D technology has proven to be effective in achieving this goal. Broken and damaged materials accumulated during construction,

such as bricks, wood, and concrete, can be repurposed using 3D printing. Additionally, industrial and construction waste can be recycled and transformed into building materials, significantly reducing waste generation. Furthermore, the use of 3D technology in construction processes releases fewer carbon emissions, contributing to environmental sustainability (Ford & Despeisse, 2016).

2. **Durability of Building Projects**: Building collapses have become a prevalent issue in many developing countries. However, 3D printing offers a solution by enabling the creation of structures capable of withstanding adverse environmental conditions like earthquakes and hurricanes. Recent discoveries show that the construction and assembly methods used in 3D printing contribute to the durability of structural elements. Enhancing building stability helps minimize fatalities resulting from collapses and increases client satisfaction (Khajavi et al., 2013a, b, c).
3. **Improved Project Management**: The integration of 3D printing technology in the construction industry has brought significant improvements in project planning, execution, and timely delivery. 3D printing resolves several challenges associated with traditional construction methods, leading to more efficient project management (Ford & Despeisse, 2016).
4. **Innovative Design Solutions**: One of the key benefits of 3D technology is its capacity for innovation. By adopting 3D printing in the built environment, architects and planners can design and construct unique and complex structures that deviate from conventional norms. 3D printing allows for the creation of custom designs and shapes, including distinctive facades and curved walls, unlike traditional rectangular structures. This ability to innovate in building design enhances client satisfaction by providing personalized and creative structures. Moreover, 3D printing offers cost advantages compared to conventional methods (Khajavi et al., 2013a, b, c).

In conclusion, 3D printing, or Additive Manufacturing, has become an indispensable technology in the built environment. Its application in civil engineering shows promise, particularly in constructing buildings with complex shapes. The benefits of 3D printing include cost reduction, increased accuracy through automation, material savings, reduced labor requirements, improved work safety, time savings, and the ability to design intricate 3D shapes (Khajavi et al., 2013a, b, c).

8.2.2 *Automated Robots*

Automation in logistics and transportation is an emerging technology that is being explored for various applications, including manufacturing plant logistics and last-mile delivery (Lever & Ray, 2008; Limosani et al., 2018). Automated robots are particularly relevant in the field of freight and logistics, and their impact on freight transportation depends on factors such as economic viability, public accessibility, stakeholder acceptance, and associated benefits. These benefits encompass economic advantages, administrative efficiency, reliability, and usability in challenging conditions. For instance, the economic benefits include cost savings through reduced labor requirements, decreased fixed production costs, and time savings (Karabegović et al., 2015; Krüger et al., 2009; Sprunk et al., 2017). While these benefits primarily focus on operational activities within the logistics industry, automated robots also streamline administrative tasks, such as using decision-making methods for parts supply operations to increase production volumes and productivity (Kousi et al., 2016). Additionally, these robots contribute to sustainable transportation solutions within warehouses (Nia et al., 2017) and offer ease of use in extreme working conditions (Lever & Ray, 2008). Logistics service providers commonly employ automated robots for short-distance deliveries, while shippers utilize them for in-house transportation of raw materials or finished products (Limosani et al., 2018; Kousi et al., 2016). Their application is especially valuable in warehouse environments for efficient short-distance deliveries (Yi et al., 2017), as it is easier to control and optimize robots within a smaller, self-organized setting. Furthermore, automated robots find utility in road construction, particularly in tasks like road grading (see Fig. 8.2).

Components of a Robot

A robotic system consists of various parts that work together to enable its functionality. Anudari (2019) outlines the following parts that make up a robot:

1. **Body:** The body of a robot serves as its physical structure, similar to the human body. It provides rigidity and protection to the internal components of the robot

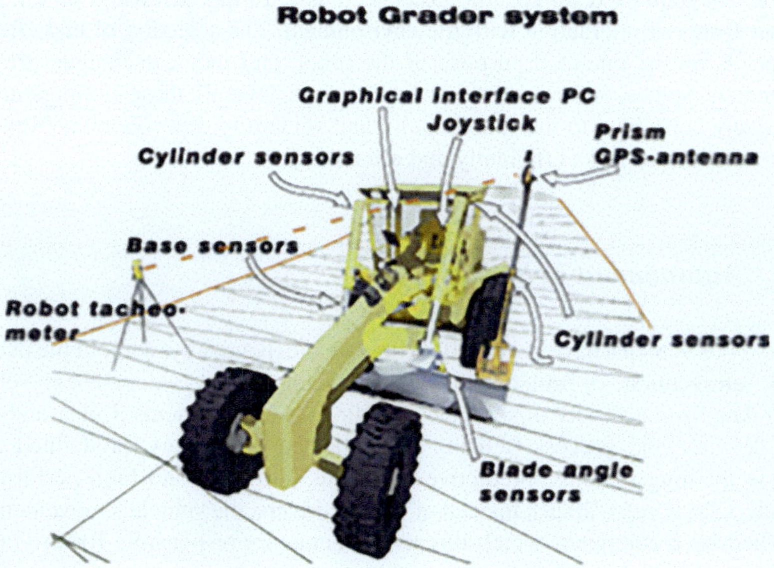

Fig. 8.2 Robot grader. 3-D blade control system principle for road grader. (Heikkilä & Jaakkola, 2002)

against external damage. Additionally, the body provides a framework for attaching important parts such as sensors. It is designed to be strong and resistant to vibrations.

2. **Locomotion**: Robots, whether fully automated or not, need the ability to move. The locomotion components enable the robot to traverse from one location to another autonomously, without requiring physical assistance from a human. Locomotion systems can take various forms, including legs for walking, wheels for driving, or wings for flying, depending on the design and intended purpose of the robot.
3. **Sensors**: Sensors play a crucial role in a robot's functionality. They detect and receive information from the environment and transmit electrical signals to the robot's control unit. Sensors are utilized to scan the surroundings, enabling the robot to avoid collisions with objects and to create maps of the task environment. They provide valuable input for the robot's decision-making and navigation.
4. **Manipulators**: Manipulators are responsible for the movement and manipulation of objects during tasks. The type and number of manipulators employed by a robot depend on the specific functions it performs, and they can vary in shape and size. Manipulators act as the robot's arms when interacting with the work environment. The simplest manipulators have three Degrees of Freedom (DOF), representing the number of axes on which they can move. For example, a manipulator with three DOF can move up and down along the x-axis, back and forth along the y-axis, and rotate around the z-axis. More complex robots may have a higher number of DOF, providing increased flexibility and dexterity.
5. **End Effectors**: End effectors are tools or devices attached to the robot's manipulator. They are used for specific tasks such as grasping, screwing, holding, and other forms of interaction with the environment. The selection of end effectors depends on the intended purpose of the robot, and they can include grippers, hammers, spikes, or screwdrivers, among others. Overall, these components collectively contribute to the functionality and versatility of a robot, enabling it to perform various tasks efficiently and effectively.

8.2.3 Autonomous Vehicles

Autonomous vehicles have the ability to operate independently, without the need for human intervention, by utilizing advanced sensing technologies. These vehicles collect data from a variety of sources, including sensors, cameras, Lidar, and ultrasonic devices, and process it using sophisticated algorithms, predefined rules, machine learning systems, predictive modeling, software, and high-performance computers (Synopsys, 2020). Instead of human drivers, the vehicle's movements are controlled by a computer, which directs the actuators responsible for driving the vehicle (see Fig. 8.3).

The advantages of autonomous vehicles encompass various aspects such as time efficiency, reduced carbon emissions for a more sustainable transport system, cost

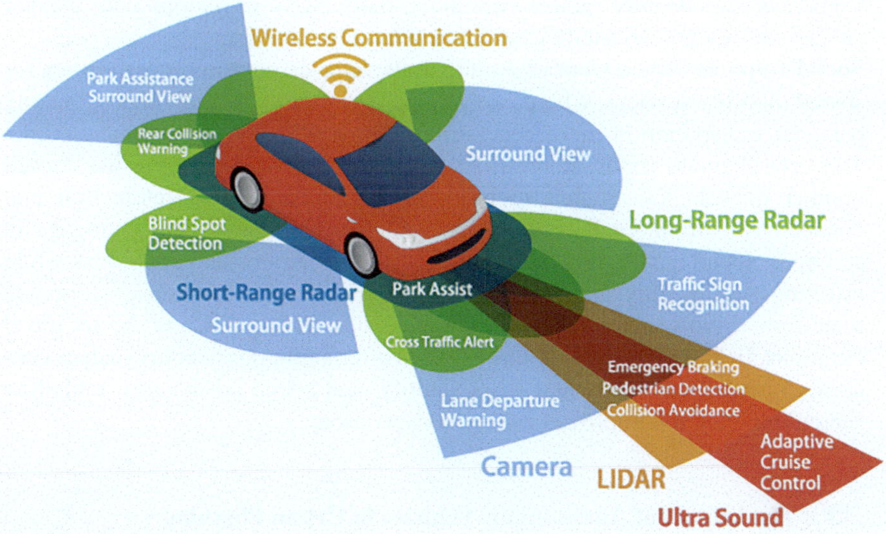

Fig. 8.3 A typical image of an autonomous vehicle and technologies in self-driving car. (Source: Olaverri-Monreal, 2020)

savings, and enhanced operational efficiency at terminals (Ross & Guhathakurta, 2017; Liang et al., 2016a, b; Huang & Zheng, 2016; Ventura et al., 2013). However, the economic benefits rely on factors like platooning dynamics and opportunities for drivers to rest during autonomous operations (Larsen et al., 2019a, b). Narayani and Krishna Kumar (2021) classified the impact of autonomous vehicles in different sectors, including:

- **Economic Impact:** Fully adopting autonomous vehicles in urban transportation planning is projected to result in approximately $800 billion in annual savings (Narayani & Krishna Kumar, 2021). These savings stem from cost reductions associated with decreased accidents, improved fuel efficiency, enhanced transportation accessibility, reduced labor costs (no drivers required), shorter travel times, minimal infrastructure expansion, decreased healthcare expenses due to accident prevention, new job opportunities, reduced accident likelihood, lower vehicle insurance premiums, and reduced cost per kilometer for consumers and businesses.
- **Environmental Impacts:** Autonomous vehicles have the potential to mitigate the environmental impact of road transportation, contingent upon the implementation strategies of stakeholders in the transport and built environment sectors. Benefits include reduced fuel consumption, lower cost per kilometer, and enhanced time efficiency, which collectively lessen the environmental footprint. Furthermore, the use of advanced powertrain technologies like Battery Electric Vehicles (BEVs) or hydrogen fuel cell vehicles (HFCVs) can contribute to improved air quality. Autonomous Electric Vehicles, with zero tailpipe emis-

sions, can significantly reduce emissions, particularly in regions with cleaner energy sources for electricity generation (Gavanas, 2019).

- **Social Impacts**: The advent of autonomous vehicles opens up possibilities for shared mobility services, where one vehicle serves multiple individuals, making personal transportation more accessible (Narayani & Krishna Kumar, 2021). This can alleviate traffic congestion on major routes by reducing the overall number of vehicles. Shared mobility also promotes social interaction and connectivity within small autonomous vehicles, enabling people to engage in activities, spend quality time with family, foster creativity, and even work while traveling. Additionally, autonomous vehicles contribute to fuel conservation due to uniform driving styles, reduced time spent at traffic signals, and shorter travel durations. In addition, autonomous vehicles offer numerous benefits such as economic savings, environmental sustainability, and social advantages, including shared mobility and improved energy conservation.

8.2.3.1 The Impact of Autonomous Vehicles in Urban Planning

Land use and transport systems are intricately linked, and the primary objective of emerging transport technologies is to alleviate congestion and enhance safe travel. One approach to achieving this goal is through urban planners integrating diverse datasets to incorporate sustainable mobility solutions and consider socio-economic and environmental objectives (Gavanas, 2019). A significant shift in planning practices focuses on mixed-use development, which reduces the need for extensive travel and promotes fuel efficiency. By providing amenities within closer proximity and reducing the number of vehicles on the road, carbon emissions and air pollution can be effectively minimized. Real-time tracking of vehicle connectivity enables the creation of efficient junctions and traffic signals to alleviate congestion, resulting in reduced fuel consumption and pollution levels. Anticipated future changes in city planning include the transformation of parking lots into green spaces and the adoption of mixed development approaches (Narayani & Krishna Kumar, 2021). Moreover, autonomous vehicles will contribute to achieving compact and dense settlements. Ultimately, the adoption of autonomous vehicles and other emerging technologies will drive a paradigm shift in city design, transitioning from a vehicle-centric approach to one that prioritizes human-centric environments.

8.2.3.2 Risks Associated with Autonomous Vehicles

Autonomous vehicles, while offering numerous benefits, also come with certain risks and challenges that need to be addressed. Some of the key risks associated with autonomous vehicles are as follows:

- Cybersecurity Risks: Autonomous vehicles are vulnerable to cyberattacks, and humans can be the primary carriers of such attacks. A single point of failure in the network can disrupt the entire system or even cause accidents. Hackers may exploit vulnerabilities to gain control over individual vehicles or compromise the entire network, leading to potential destruction and safety hazards (Gerdes et al., 2016; Koopman & Wagner, 2017).

- Coexistence with Conventional Vehicles: The coexistence of autonomous vehicles with traditional human-driven vehicles on the same road poses risks and challenges. The differences in driving behavior and capabilities between autonomous and human-driven vehicles can potentially result in accidents or unsafe situations (Cao et al., 2020).
- Infrastructure Challenges: The implementation of autonomous vehicles, with their increased time efficiency and reduced travel space requirements, may impact the need for future infrastructure upgrades. Stakeholders may be reluctant to invest in road infrastructure improvements due to the perception of reduced capacity utilization, which can hinder the development of infrastructure to support the safe and efficient operation of autonomous vehicles (Moylan et al., 2020).

It is crucial for stakeholders, including policymakers, manufacturers, and cybersecurity experts, to address these risks through robust cybersecurity measures, effective regulations, and infrastructure planning that considers the integration of autonomous vehicles into the existing transportation ecosystem (Koopman & Wagner, 2017; Gerdes et al., 2016). By proactively managing these risks, the full potential of autonomous vehicles can be realized while ensuring safety and security on the roads.

8.2.4 Blockchain

Transport can benefit greatly from the utilization of blockchain technology, which has been specifically developed for the transport sector. Blockchain serves as a method of record-keeping, where immutable records of transactions are accessible to all participants. Despite facing challenges such as standardization in record-keeping methods and industry-wide acceptance, experts widely agree that blockchain will play a pivotal role in fleet-management solutions. Its usefulness extends to creating transparent and automated payment systems, simplifying transactions for businesses, vendors, suppliers, owners, and fleet managers. The advantages associated with blockchain in the transport sector include track and trace capabilities, credit evaluation, enhanced transportation visibility, strengthened security, improved data sharing, addressing provenance issues, and ownership registry (Chaudhuri et al., 2018; Li et al., 2018; Wu et al., 2017; Polim et al., 2017; Akram & Bross, 2018). In logistics and intelligent order delivery tracking, blockchain proves particularly valuable due to the increasing demand for same-day and express delivery services, surpassing the capabilities of traditional tracking technologies. Its scalable and immediate solution for order tracking and authentication enhances the quality of services provided by transport and delivery companies. Additionally, blockchain serves as a reliable tool for data verification within the transport and logistics sector, aiding in tracking truck fleets, routes, traffic incidents, and related operations. It contributes to greater transparency, security, and the minimization of potential fraud or threats.

Walburg (2021) identified several benefits of blockchain technology for transportation and these include:

- **Increased security** – Firstly, it offers increased security by preventing fraud and attacks on fleets through data signature alteration, facilitating the identification of fraudulent activities and their perpetrators. Moreover, blockchain's resistance to retrospective data modification ensures the integrity of the stored information, making it suitable for monitoring mechanisms in transport and logistics.
- **Better Traceability** – Blockchain provides better traceability by effectively tracking updates throughout the value chain and all processes involved, offering insights into the product's lifecycle stages such as production, delivery, logistics, and distribution. This precise tracking capability aids in fraud prevention, accurate origin checking, and ensures the overall quality of the product and transport service.
- **Time reduction** – It enables time reduction through faster payments and transaction settlements, significantly reducing waiting times by eliminating lengthy verification and billing processes. All parties have access to a single version of the agreed share ledger data, streamlining financial operations.
- **Transparency** – It also offers transparency as all transactions are recorded in an immutable manner, and changes made to public blockchains are visible to all participants, fostering greater openness among involved parties.

One notable example of blockchain application in the transport sector is IBM's collaboration with other companies to develop the first blockchain-based automotive trading platform for mobility services. This technology enables secure and transparent payments within vehicles, allowing drivers to conveniently pay for tolls, parking, and car-sharing, or receive payments for car-sharing services. An essential aspect of this project involves the development of a reliable and autonomous agent that enables cars to react to their surroundings, navigate routes, and perform diagnostic checks without human intervention. Although blockchain technology is still evolving, particularly within the transport sector, stakeholders have already begun implementing or developing blockchain-based solutions. The adoption of this technology may require time, but the growing popularity and profitability of implementing appropriate software suggest that more companies will move in this direction.

8.2.5 Vehicle-to-Vehicle Technology (V2V)

This refers to the elements comprising dynamic traffic data acquisition and distribution systems. It encompasses the communication infrastructure that facilitates message dissemination among vehicles, typically involving the hardware components of information transmission technologies. According to Jin et al. (2013), vehicle data and information can be obtained through non-vehicle-embedded technologies like GPS, which enables the identification of vehicle speed and location. Alternatively, this data can be derived from vehicle-based sensors, where the vehicle's computer provides information on speed and location. By incorporating additional data such as latitude and longitude, a more comprehensive understanding of the surrounding vehicle positions can be obtained. This information proves valuable in accident prevention and theft detection through effective monitoring of vehicles in logistics and transportation operations.

8.2.6 *Vehicle-to-Infrastructure Technology (V2I)*

Vehicle-to-Infrastructure (V2I) technology encompasses the framework that enables seamless communication between vehicles and roadway infrastructure. In sophisticated implementations, crucial vehicle data such as speed and location can be transmitted to a centralized server. This server acts as a repository, collecting and consolidating speed and location information from all vehicles, which can then be utilized for diverse connected vehicle applications such as route planning, incident identification, and safety and mobility enhancements. The hardware elements involved in V2I technology include roadside sensors and DSRC antennas (Jin et al., 2013).

8.2.7 *Drone*

Drones are flying robots that can be remotely controlled using flight software and Global Positioning Systems (GPS). They serve as aerial vehicles for data collection purposes. Drones come in various sizes and are equipped with topographic measurement capabilities, which make them suitable for surveying buildings, roads, highways, bridges, and other surface areas. One of the key advantages of drones is their ability to access remote or inaccessible areas, such as places affected by disasters. This accessibility is particularly advantageous in drone deliveries (Chowdhury et al., 2017). Drones offer benefits over traditional transportation methods, including faster delivery times compared to conventional means and reduced transportation costs (Rabta et al., 2018; Ha et al., 2018; Dong et al., 2021a, b). These benefits, such as speedy delivery and cost reduction, are especially effective for last-mile delivery (Rao et al., 2016; Rabta et al., 2018; Ha et al., 2018). Furthermore, Chiang et al. (2019) suggest that drones are highly cost-effective and environmentally sustainable when optimized route planning strategies are implemented.

8.2.8 *Cloud Computing and Crowdsourcing*

Cloud computing refers to the provision of IT infrastructure, platforms, and software over the Internet, enabling end-users to remotely access powerful computing resources through broadband connections. Crowdsourcing combines the concepts of the wisdom of crowds and outsourcing of information. It involves leveraging the collective intelligence of the public to perform business-related tasks that an organization would typically handle internally or outsource to a third party. The integration of cloud computing and crowdsourcing allows transportation service providers to efficiently harness the knowledge of a large crowd by removing barriers to information sharing, data transmission, organization, and analysis. Through mobile communication technologies, individual drivers can contribute their inputs and share their available travel data, which can then be collected and communicated to a processing server.

8.2.9 *Electric Vehicles*

Previously, Electric vehicles (EVs) were not considered cost-effective compared to diesel vehicles (Lebeau et al., 2015a, b). However, advancements in technology and analysis methods have refined this perspective. The use of EVs has been better assessed through lifecycle analysis of greenhouse gas (GHG) emissions and total cost of ownership (Dong et al., 2021a, b).

The use of fossil fuel-powered vehicles in transportation negatively impacts local air quality and the climate. Exhaust emissions from these vehicles, such as CO_2, nitrogen oxides, and particulate matter, contribute to air pollution. Additionally, road traffic is the primary source of noise and CO_2 emissions in the environment (Jóźwicka, 2021a, b, c). Introducing electric vehicles into the fleet can significantly reduce overall GHG emissions and air pollution, especially if the electricity used is from renewable sources. Even when electricity comes from fossil fuels, a transition to electric vehicles can still improve local air quality and reduce noise levels.

Boren (2018a, b) explains that the main difference between an EV and an Internal Combustion Engine Vehicle (ICEV) lies in the driveline from refueling/charging to the driveshaft. ICEVs use fuel pumps to fill the tank with gasoline or diesel, which is combusted to power the engine, driveshaft, gearbox, and wheels. In contrast, EVs charge their batteries, which then power the electric motor and rotate the wheels. The efficiency of an electric motor is around 85–90%, while a combustion engine's efficiency ranges from 35% to 45% depending on fuel and technology. However, the electric driveline incurs losses during battery charging and discharge over time. An electric car utilizes only about 25% of the energy required by a gasoline-powered car to transport one person for 100 km. Currently, EVs offer advantages in terms of efficiency and emissions during the use phase compared to fossil fuel-powered vehicles and diesel. Electric cars powered by renewable energy sources can reduce

GHG emissions by 90% compared to gasoline-powered cars during the use phase (Boren, 2018a, b). To summarize, Polzin (2016a, b, c) highlights the role of emerging technologies in urban transportation and mobility in the following areas:

- **Safety:** The safety of travelers has always been a concern, and emerging technologies can address personal safety, security, and protection from accidents. Automated vehicles are expected to reduce accidents caused by human error. Real-time information on accident causes and modes of transport can help passengers make informed decisions about route selection and avoid certain modes of transport. Ensuring that automated vehicles cannot be hijacked for malicious purposes is also crucial.
- **Economic Impact:** Technology is predicted to significantly impact travel costs in terms of time and money. It can transform the pricing structure for transportation, potentially reducing the high fixed costs associated with vehicle ownership. Self-driving fleets, for example, are projected to be more cost-effective than private car ownership. Shared autonomous trips to similar destinations could provide mobility costs as low as $0.20 to $0.50 per passenger mile. These cost differences are expected to bring about transformative effects in automated mobility, particularly in urban areas where auto ownership costs are highest and transit use is most productive (Mims, 2016a, b).

Convenience, Reliability, Comfort, and Flexibility Associated with Automated or Mobility as a Service Opportunity:

The consumption of any product is influenced by various factors, including consumer awareness and convenience (Aderibigbe & Gumbo, 2023). The advent of the internet has made online shopping more convenient and accessible, allowing travelers to have convenient access to information about travel options through web and smartphone-based applications. These technological advancements have provided travelers with the ability to understand choices, plan itineraries, make payments, and provide feedback, breaking down barriers of market presence, physical mode options, and traditional marketing. Furthermore, emerging technologies have facilitated intermodal integration and coordination of services across different jurisdictions, resulting in enhanced convenience (Polzin, 2016a, b, c). Cross-geography standardization of applications, such as OneBusAway and Uber, has contributed to the convenience of travelers across communities.

In addition to convenience, technology plays a crucial role in improving the reliability of both private and public vehicles. Real-time monitoring of vehicle systems, improved passenger safety features, and in-vehicle communication/entertainment opportunities contribute to enhanced service reliability (Polzin, 2016a, b, c). Technological features also have the potential to resolve issues related to the overall travel experience and the attractiveness of different travel options, thus influencing travelers' willingness to transition from their current modes of travel (Rayle et al., 2016).

- **Environmental and Social Impact of Technology on Travel Behavior:** The environmental and social impact of technology on travel behavior is a significant aspect to consider (Aderibigbe & Gumbo, 2023). People's travel decisions are influenced by their perceptions of the social and environmental consequences associated with different means of travel and their subsequent effects on individuals and the community. The cultural significance of automobiles as a status symbol plays a significant role in the adoption of alternative modes of travel. Private cars offer various benefits such as comfort, personal space, storage capacity, and the ability to customize features like temperature and audio settings according to individual preferences. Moreover, individuals can modify their safety risks through vehicle selection and operating decisions. These factors make it challenging for people to transition to alternative modes of transport, such as shared vehicles and public transportation, as they may not provide the necessary privacy. Shared or automated modes of transport often limit certain opportunities associated with private car use. Vehicle selection is often associated with status and personal values, including considerations of luxury, sportiness, energy efficiency, and safety. It is commonly assumed that automated vehicles are inherently energy efficient and environmentally friendly due to factors like electrical propulsion systems, optimized driving behaviors, travel path optimization, reduced vehicle weight, and the potential for convenient vehicle sharing and higher occupancy rates, leading to optimized roadway flow patterns. The integration of mobility-as-a-service and automation technologies in public transportation will have significant implications for urban transit travel, attracting passengers through real-time access to travel information such as departure and arrival times and the nearest bus stops (Polzin, 2016a, b, c; Rayle et al., 2016).

- **The role of technology in intermodal and interjurisdictional service coordination:** Technology plays a crucial role in facilitating intermodal and interjurisdictional service coordination, benefiting passengers through the provision of various information systems (Krajzewicz et al., 2012). These technologies include customer information systems, travel scheduling systems, and fare payment systems, which collectively enhance public transit services by enabling digital information dissemination and coordination. In the past, the physical presence of transit facilities and the standardization of technologies across different systems, along with regionalization of service provision, have been instrumental in creating awareness among travelers and offering convenient access to a network of transit services (Cats et al., 2017). However, advancements in technology have now enabled the use of digital platforms to comprehensively understand available service options, providing convenient information and facilitating fare payments. Consequently, there is a reduced need for extensive standardization or consolidation of transit systems. The reliance on physical rail availability and service coverage is diminished, as smartphone applications can provide sufficient information for travelers to plan multimodal trips (Derrible & Kennedy, 2018). This shift reduces the bias towards service standardization and regionalization, allowing for greater flexibility and customization of operations. Furthermore, technology empowers infrequent travelers by enabling them to easily access information about available transit options, removing barriers related to unfamiliarity and complexity, and encouraging them to embark on multimodal or multi-vehicle journeys that might have seemed daunting or inconvenient in the past.

- **The role of Technology in the quality and cost-effectiveness of demand-responsive services:** Demand-responsive services, such as paratransit, face operational challenges that contribute to higher operating expenditures per passenger trip, primarily due to lower demand density and the requirement of providing curb-to-curb services and special accommodations for mobility aids (Dablanc et al., 2020). However, technology plays a pivotal role in enhancing the quality and cost-effectiveness of these services. By leveraging technology, logistics for demand-responsive services can be significantly improved, opening up possibilities for multiple service providers to cater to diverse travel markets (Yang et al., 2018). This allows for a better alignment of economies of scale, vehicle types, and service providers to efficiently meet the unique needs of different travelers. For instance, many transportation agencies operate a large fleet of vehicles equipped to handle wheelchairs and mobility aids. With technology, it becomes feasible to designate these vehicles for travelers requiring such accommodations, while utilizing smaller, more suitable vehicles for others. This approach optimizes the deployment of vehicles, matching them precisely with the specific needs of travelers, and facilitating coordinated service provision (Hickman et al., 2020). As a result, logistics efficiency is improved, and trip scheduling and assignment can be optimized, leading to enhanced quality and cost-effectiveness of demand-responsive services.
- **The effect of technologies on long-range planning and infrastructure investment:** Long-range planning and capital investment decision-making in transit face a crucial challenge in light of emerging technologies and the array of travel options they offer. This raises the question of how these planning processes should be influenced by the presence of these technologies. In particular, many communities are currently engaged in planning major capital investments in fixed-guideway public transit systems, and the success of these investments may be affected by the availability of alternative mobility options (Cervero & Duncan, 2021). It is essential to carefully consider these issues due to the substantial costs involved and the long-term nature of these infrastructure assets. These fixed-guideway commitments may still have a considerable economic lifespan when new travel options emerge, potentially leading to market competition and diminishing the envisioned productivity of the investments made during the planning phase.

8.3 Challenges of Adopting Emerging Technologies in Urban Transportation Planning

The widespread adoption of emerging technologies in urban transportation planning faces various barriers and obstacles. One of the main challenges is related to the implementation of personal mobility solutions based on electric vehicles (EVs) and buses. High prices of EVs and the use of scarce metals in batteries have been identified as significant barriers to the broader acceptance of electromobility (Jóźwicka, 2021a, b, c). In addition, improvements in EVs are needed, such as increasing their driving range and reducing charging times. Currently, the fastest charging stations take around 20–30 min to charge a vehicle for a 100 km drive. To fully leverage the benefits of electromobility, an enhanced infrastructure is required, including widespread public charging points comparable to conventional fueling stations, as well as an expansion of renewable energy generation capacity. It is worth noting that electric vehicles are also generally more expensive than conventional vehicles.

In the realm of transportation, four emerging technologies have been identified as significant: Big Data Analytics, Artificial Intelligence (AI), Blockchain, and the Internet of Things (IoT). These technologies rely on data as their foundation and have the potential to support new models, algorithms, and applications, addressing challenging problems from a human perspective. However, effectively obtaining and understanding knowledge from data will continue to pose challenges in the coming years. For example, IoT relies on RFID tags to collect, process, and transmit data, serving as the interface between data and real transportation activities. While IoT has the potential to be integrated into various physical movements, the cost barrier associated with commercial usage remains a challenge, particularly in the low-margin logistics industry. On the other hand, Blockchain offers a novel approach to creating trust among data-sharing actors, leveraging its simple algorithms to ensure data immutability. This decentralized trust, without the need for intermediaries like banks or brokers, is particularly relevant in the transportation sector where freight flows involve stakeholders with diverse backgrounds across the globe (Kumar et al., 2019a, b; Dobrovnik et al., 2019a, b).

Furthermore, emerging technologies such as 3D Printing, Automated Robots, Autonomous Vehicles, and Drones play a role in new Automation Systems for freight transportation. These technologies aim to enhance the efficiency of freight operations; however, their application areas are currently limited and complementary to existing solutions. Technical and economic constraints still restrict the range of products that can be 3D printed. Automated robots and drones are often employed in combination with traditional road transportation for last-mile deliveries. Autonomous vehicles are being increasingly utilized in confined areas where it is legally permissible, such as mines, terminal areas, and seaports, while other automated systems and vehicles, like platooning, are undergoing testing for long-distance transport (Halassi et al., 2019a, b; Sacramento et al., 2019a, b; Aurambout et al., 2019a, b; Liang et al., 2016a, b; Larsen et al., 2019a, b). In conclusion, emerging technologies, including drones, 3D printing, and electric vehicles, play a crucial role in urban transportation planning by reducing negative externalities associated with transport. However, the adoption of these technologies in planning processes remains low, particularly in cities of developing countries. Therefore, there is a need for increased orientation and awareness to make these technologies more attractive so as to enable it address the transportation challenges faced worldwide.

References

Aderibigbe, O.-O., & Gumbo, T. (2023). Geospatial analysis of telecommunication and travel behaviour of households: Lessons from a developing city in Nigeria. *Journal of Sustainable Development of Transport and Logistics, 8*(2), 71–85. https://doi.org/10.14254/jsdtl.2023.8-2.5.

Akram, A., & Bross, P. (2018). Trust, privacy and transparency with blockhain technology in logistics. The 12th Mediterranean Conference on Information Systems (MCIS), Corfu, Greece, 2018

Ankner, B. (2014). *3D printers: Changing transportation as we know it.* Accessed on January, 2023 from https://www.enotrans.org/article/3d-printers-changing-transportation-know/

Anudari N. (2019). *Robotics in construction of concrete and masonry structures.* A thesis submitted in partial fulfilment for the degree of Bachelor in Civil Engineering branch of Building Structures 2019 Czech Technical University in Prague Faculty of Civil Engineering.

Aurambout, J.-P., Gkoumas, K., & Ciuffo, B. (2019a). Last mile delivery by drones: an estimation of viable market potential and access to citizens across European cities. *European Transport Research Review, 11*(1), 30.

Aurambout, J., Deuse, J., & Kreutzberger, E. (2019b). An automated last-mile delivery concept using unmanned aerial vehicles and automated road vehicles. *Procedia Computer Science, 151*, 1–8. https://doi.org/10.1016/j.procs.2019.05.074

Boren, J. (2018a). Environmental and energy impacts of electric vehicles: Literature review and synthesis. *Journal of Industrial Ecology, 22*(5), 929–946.

Boren, V. (2018b). *Towards sustainable personal mobility with electric cars and buses.* A Doctoral Dissertation in Strategic Sustainable Development, Submitted to Department of Strategic Sustainable Development, Blekinge Institute of Technology SWEDEN. Accessed January 11, 2020 from https://www.diva-portal.org/smash/get/diva2:1245625/FULLTEXT02.pdf

Briggs, M., & Sundaram, K. (2016a). *Environmentally sustainable innovation in automotive manufacturing and urban mobility* (36 p). Frost & Sullivan.

Briggs, R., & Sundaram, A. (2016b). *Mobility-as-a-service: A roadmap for the future.* Deloitte University Press.

Cao, D., Wei, T., Ma, X., Li, K., & Zhang, C. (2020). Autonomous vehicle and human-driven vehicle coexistence based on cooperative lane changing. *IEEE Access, 8*, 16845–16855.

Cats, O., Jenelius, E., & Bierlaire, M. (2017). How smart growth and accessibility favour walking, cycling and public transport in dense cities. *International Journal of Sustainable Transportation, 11*(1), 1–11. https://doi.org/10.1080/15568318.2015.1104443

Cervero, R., & Duncan, M. (2021). Urban transit systems and technology. In *A research agenda for transport planning and policy in the global south* (pp. 55–71). Edward Elgar Publishing. https://doi.org/10.4337/9781789903297.00012

Chaudhuri, A., Dukovska-Popovska, I., Subramanian, N., Chan, H. K., & Bai, R. (2018). Decision-making in cold chain logistics using data analytics: A literature review. *The International Journal of Logistics Management, 29*(3), 839–861.

Chiang, W.-C., Li, Y., Shang, J., & Urban, T. L. (2019). Impact of drone delivery on sustainability and cost: Realizing the UAV potential through vehicle routing optimization. *Applied Energy, 242*, 1164–1175.

Chowdhury, S., Emelogu, A., Marufuzzaman, M., Nurre, S. G., & Bian, L. (2017). Drones for disaster response and relief operations: A continuous approximation model. *International Journal of Production Economics, 188*, 167–184.

Dablanc, L., Ettema, D., & Hickman, M. (2020). Paratransit in Europe and North America: The quest for efficiency. *Journal of Transport Geography, 87*, 102814. https://doi.org/10.1016/j.jtrangeo.2020.102814

Derrible, S., & Kennedy, C. (2018). The complexity and robustness of metro networks. *PLoS One, 13*(2), e0192364. https://doi.org/10.1371/journal.pone.0192364

Dobrovnik, M., Balakrishnan, H., Čelan, M., & Shyrokau, B. (2019a). Blockchain and IoT for trusted freight transportation systems. *Journal of Cleaner Production, 213*, 151–162. https://doi.org/10.1016/j.jclepro.2018.12.245

Dobrovnik, M., Herold, D., Fürst, E., & Kummer, S. (2019b). Blockchain for and in logistics: what to adopt and where to start. *Logistics, 2*(3), 18.

Dong, C., Akram, A., Andersson, D., Arnäs, P.-O. & Stefansson, G. (2021a). The impact of emerging and disruptive technologies on freight transportation in the digital era: current state and future trends. *The International Journal of Logistics Management*, [Online] Emerald *32*(2), 386–412. https://doi.org/10.1108/IJLM-01-2020-0043

Dong, T., Guan, X., Zhang, L., Luo, Y., & Zhang, Y. (2021b). The analysis of environmental life cycle assessment and total cost of ownership for electric vehicle adoption. *Journal of Cleaner Production, 289*, 125795.

Ford, S., & Despeisse, M. (2016). Additive manufacturing and sustainability: An exploratory study of the advantages and challenges. *Journal of Cleaner Production, 137*, 1573–1587.

Franckx, L. (2015a). Mobility as a service (MaaS): Does public transport have a future role in rural areas? *Transportation Research Procedia, 12*, 94–103.

Franckx, L., (2015b). *Future trends in mobility: Challenges for transport planning tools and related decision-making on mobility product and service development.* MIND-sets Deliverable no. 3.3. 49 p.

Gavanas, N. (2019). Autonomous road vehicles: Challenges for urban planning in European cities. *Urban Science* [Online] *13*(9). Available from: https://www.mdpi.com/2413-8851/3/2/61. Accessed 15 Dec 2021.

Gerdes, J. C., Hoffmann, G. M., & Zhou, H. (2016). Cybersecurity for connected and autonomous vehicles. In *Connected and autonomous vehicles* (pp. 61–76). Springer.

Giacomelli, P., & Smedberg, A. (2014). The eve of 3D printing in telemedicine: State of the art and future challenges.

Ha, Q. M., Deville, Y., Pham, Q. D., & Hà, M. H. (2018). On the min-cost traveling salesman problem with drone. *Transportation Research Part C: Emerging Technologies, 86*, 597–621.

Halassi, A. K., Al Mannai, L., & Khan, F. A. (2019a). 3D printing in the supply chain: A comprehensive review and future directions. *Technological Forecasting and Social Change, 146*, 1–13. https://doi.org/10.1016/j.techfore.2019.06.004

Halassi, S., Semeijn, J., & Kiratli, N. (2019b). From consumer to prosumer: a supply chain revolution in 3D printing. *International Journal of Physical Distribution & Logistics Management, 49*(2), 200–216.

Heikkilä, R., & Jaakkola, M. (2002). The efficiency of a 3-D blade control system in the construction of structure layers by road grader-automated design-build of road construction in Finland. In *Proceedings of the 19th ISARC 2002, 23–23 Sept. 2002.*

Hickman, M., Ettema, D., & Dablanc, L. (2020). The case for an open Paratransit architecture. *Journal of Transport Geography, 82*, 102576. https://doi.org/10.1016/j.jtrangeo.2019.102576

Huang, Q., & Zheng, G. (2016). Route Optimization for Autonomous Container Truck Based on Rolling Window. *International Journal of Advanced Robotic Systems, 13*(3), 112.

Jin, P.J., Fagnant, D., Hall, A., Walton, C.M., Hockenyos, J. & Krusee, M. (2013). *Developing emerging transportation technologies in Texas*. A technical report by Center for Transportation Research, University of Texas Austin. Accessed on January 15 2023, from https://library.ctr.utexas.edu/ctr-publications/0-6803-1.pdf

Jóźwicka, E. (2021a). Electric vehicles – Current barriers and challenges for their wider implementation. *European Transport Research Review, 13*(1), 1–14. https://doi.org/10.1186/s12544-021-00497-5

Jóźwicka, M. (2021b). A review of electric vehicle impacts on urban air quality. *Sustainability, 13*(5), 2575.

Jóźwicka, M. (2021c). *Electric vehicles: Moving towards a sustainable mobility system.* Accessed on January 2021 from https://www.eea.europa.eu/articles/electric-vehicles

Karabegović, I., Karabegović, E., Mahmić, M., & Husak, E. (2015). The application of service robots for logistics in manufacturing processes. *Advances in Production Engineering & Management, 10*(4), 185–194.

Khajavi, S. H., Gu, P., Hao, L., & Feng, Y. (2013a). A critical review of 3D printed customized construction: Opportunities, challenges, and building regulations. *Automation in Construction, 35*, 207–221.

Khajavi, S. H., Kalhori, V., Ghobadian, B., Mamat, R., & Yusof, A. M. (2013b). A review of 3D concrete printing systems and materials properties. *Structural Survey, 31*(3), 241–259.

Khajavi, S. H., Partanen, J., & Holmström, J. (2013c). Additive manufacturing in the spare parts supply chain. *Computers in Industry, 65*(1), 50–63.

Koopman, P., & Wagner, M. (2017). Challenges in autonomous vehicle cybersecurity. *ACM Transactions on Cyber-Physical Systems, 1*(1), 2.

Kousi, N., Michalos, G., Makris, S., & Chryssolouris, G. (2016). Short-term planning for part supply in assembly lines using mobile robots. *Procedia CIRP, 44*, 371–376.

Krajzewicz, D., Erdmann, J., Behrisch, M., & Bieker, L. (2012). Recent development and applications of SUMO—Simulation of urban mobility. *International Journal on Advances in Systems and Measurements, 5*(3&4), 128–138. https://doi.org/10.1109/ICC.2012.6364099

Krüger, J., Lien, T. K., & Verl, A. (2009). Cooperation of human and machines in assembly lines. *CIRP Annals, 58*(2), 628–646.

Kumar, A., Liu, R., & Shan, Z. (2019a). Is blockchain a silver bullet for supply chain management? Technical challenges and research opportunities. *Decision Sciences, 51*(1), 8–37.

Kumar, R., Singh, S. K., & Al-Turjman, F. (2019b). Blockchain technology for IoT-based transportation systems: A comprehensive survey. *IEEE Transactions on Intelligent Transportation Systems, 20*(2), 673–686. https://doi.org/10.1109/TITS.2018.2838915

Larsen, J., Davidovic, N., & Reed, T. (2019a). Fleet deployment optimization of autonomous vehicles in logistics. *Transportation Research Part B: Methodological, 121*, 173–196. https://doi.org/10.1016/j.trb.2019.01.001

Larsen, R., Rich, J., & Rasmussen, T. K. (2019b). Hub-based truck platooning: Potentials and profitability. *Transportation Research Part E: Logistics and Transportation Review, 127*, 249–264.

Lebeau, K., Brand, C., & van Schaik, A. (2015a). Will the costs of electric vehicles reduce? An analysis of total cost of ownership and its sensitivity to cost of driving and firm fuel efficiency. *Transportation Research Part A: Policy and Practice, 82*, 1–16.

Lebeau, P., De Cauwer, C., Van Mierlo, J., Macharis, C., Verbeke, W., & Coosemans, T. (2015b). Conventional, hybrid, or electric vehicles: Which technology for an urban distribution center? *The Scientific World Journal, 2015*, 1–11.

Lever, J. H., & Ray, L. E. (2008). Revised solar-power budget for Cool Robot polar science campaigns. *Cold Regions Science and Technology, 52*(2), 177–190.

Li, L., Liu, J., Cheng, L., Qiu, S., Wang, W., Zhang, X., & Zhang, Z. (2018). Credit coin: A privacy-preserving blockchain-based incentive announcement network for 35 communications of smart vehicles. *IEEE Transactions on Intelligent Transportation Systems, 19*(7), 2204–2220.

Liang, M., Su, X., Zhang, Y., & Luo, B. (2016a). A review on platooning control for intelligent transportation systems. *IEEE Transactions on Intelligent Transportation Systems, 17*(12), 3509–3521. https://doi.org/10.1109/TITS.2016.2563312

Liang, X., Pérez, B. O., Dey, S. S., Haney, H., & Kim, J. Y. (2016b). Big-data analytics drives parking policy: Evaluating adherence to meter time limits in Washington, D.C. *Transportation Research Record, 2594*(1), 107–117.

Limosani, R., Esposito, R., Manzi, A., Teti, G., Cavallo, F., & Dario, P. (2018). Robotic delivery service in combined outdoor–indoor environments: technical analysis and user evaluation. *Robotics and Autonomous Systems, 103*, 56–67.

Mims, C. (2016a, June 20). Driverless cars to fuel suburban sprawl: Research suggests advances in transportation technology contribute to bigger and more sprawling cities. *Wall Street Journal.*

Mims, C. (2016b). *The environmental and economic impacts of autonomous vehicles*. Center for Transportation Studies, University of Minnesota.

Morawski, M., Glińska, E., & Jach, K. (2019). Determinants of mobile application users' behavior—A research model. *Sustainability, 11*(17), 4752.

Moylan, E., Hall, C., & Hesse, M. (2020). The Future of automated mobility: Scenarios for Canada. Canadian Transportation Research Forum.

Narayani, A.R. & Krishna Kumar, K. (2021). *IOP conference series: Materials science and engineering* 1130 012037

Nia, A. R., Haleh, H., & Saghaei, A. (2017). Dual command cycle dynamic sequencing method to consider GHG efficiency in unit-load multiple-rack automated storage and retrieval systems. *Computers & Industrial Engineering, 111*, 89–108.

Olaverri-Monreal, C. (2020). *Autonomous vehicles and smart mobility related technologies* [Online]. Available from: https://www.infocommunications.hu/documents/169298/2607207/InfocommJ_2016_2_3_Olaverri.pdf. Accessed 20 Nov 2020

Polim, R., Hu, Q. & Kumara, S. (2017), *Blockchain in megacity logistics*. Proceedings of the 2017 industrial and systems engineering conference, p. 7.

Polzin, S. (2016a). Emerging technologies and their impact on mobility. In *Research on transportation systems planning and management* (Advances in Intelligent Systems and Computing, 416) (pp. 445–460). https://doi.org/10.1007/978-3-319-29175-2_38

Polzin, S. (2016b). *Automated vehicles and emerging mobility options: Policy and planning strategies for the transportation renaissance*. Florida State University, College of Engineering.

Polzin, S. E. (2016c). *Implications to public transport of emerging technologies*. A report for the National Center for Transit Research. Accessed on December 5 2020 from https://www.nctr.usf.edu/wp-content/uploads/2016/11/Implications-for-Public-Transit-of-Emerging-Technologies-11-1-16.pdf

Rabta, B., Wankmüller, C., & Reiner, G. (2018). A drone fleet model for last-mile distribution in disaster relief operations. *International Journal of Disaster Risk Reduction, 28*, 107–112.

Rao, B., Gopi, A. G., & Maione, R. (2016). The societal impact of commercial drones. *Technology in Society, 45*, 83–90.

Rauch, C. (2017a). Die Evolution der Mobilität. ADAC e.V. Germany. 50 p.

Rauch, E. (2017b). *The future of mobility – How digital networking is transforming the transportation sector*. SAP SE.

Rayle, L., Dai, D., Chan, N., Cervero, R., & Shaheen, S. (2016). Just a better taxi? A survey-based comparison of taxis, transit, and ridesourcing services in San Francisco. *Transport Policy, 45*, 168–178. https://doi.org/10.1016/j.tranpol.2015.09.009

Ross, C., & Guhathakurta, S. (2017). Autonomous vehicles and energy impacts: A scenario analysis. *Energy Procedia, 143*, 47–52.

Sacramento, C., Bogusz, G., Ferreira, B., & Olszewski, P. (2019a). Review of robotic systems for last-mile delivery. *IFAC-PapersOnLine, 52*(30), 276–281. https://doi.org/10.1016/j.ifacol.2019.12.478

Sacramento, D., Pisinger, D., & Ropke, S. (2019b). An adaptive large neighborhood search metaheuristic for the vehicle routing problem with drones. *Transportation Research Part C: Emerging Technologies, 102*, 289–315.

Sprunk, C., Lau, B., Pfaff, P., & Burgard, W. (2017). An accurate and efficient navigation system for omnidirectional robots in industrial environments. *Autonomous Robots, 41*(2), 473–493.

Sułkowski, Ł., Kolasińska-Morawska, K., & Morawski, P. (2017a). The Internet of Things – A physical logical and business model. *International Journal of Contemporary Management.*

Sułkowski, Ł., Turek, A., & Jagodziński, P. (2017b). Internet of Things (IoT) in supply chain management. In T. A. Grzeszczyk & M. Abram (Eds.), *Advances in intelligent systems and computing* (Vol. 575, pp. 309–317). Springer.

Synopsys, I.N.C. (2020). *What is an autonomous car? – How self-driving cars work* [Online]. Available from: https://www.synopsys.com/what-is-autonomous-car.html. Accessed 14 Nov 2020.

U.S. Department of Transportation, Federal Highway Administration. (2013a). *Impacts of technology advancements on transportation management center operation*. Accessed on January, 2023 from https://ops.fhwa.dot.gov/publications/fhwahop13008/fhwahop13008.pdf

U.S. Department of Transportation, Federal Highway Administration. (2013b). *The Intelligent Transportation Systems (ITS) professional capacity building program*. Retrieved from https://ops.fhwa.dot.gov/publications/fhwahop13037/chapter2.htm

Ventura, J., Valdebenito, V., & Golany, B. (2013). A dynamic inventory model with supplier selection in a serial supply chain structure. *European Journal of Operational Research, 230*(2), 258–271.

Verhoef, L. A., Budde, B. W., Chockalingam, C., García Nodar, B., & van Wijk, A. J. M. (2018). The effect of additive manufacturing on global energy demand: An assessment using a bottom-up approach. *Energy Policy, 112*, 349–360.

Walburg, M. (2021). *What is blockchain in transportation? How is it used*. Accessed on 12 December, 2022 from https://binarapps.com/what-is-blockchain-in-transportation-how-is-it-used/#:~:text=Blockchain%2C%20therefore%2C%20provides%20a%20reliable,of%20potential%20fraud%20or%20threats

Wu, H., Li, Z., King, B., Ben Miled, Z., Wassick, J., & Tazelaar, J. (2017). A distributed ledger for supply chain physical distribution visibility. *Information, 8*(4), 137.

Yang, J., Hall, R. W., & Kuby, M. (2018). Optimization models for ridesharing: A review. *Networks and Spatial Economics, 18*(3), 377–404. https://doi.org/10.1007/s11067-017-9364-8

Yi, Z., Nianjun, L., & Jun, C. (2017). A automatic matching algorithm and simulation for irregular cigarette package stacking. *Procedia Engineering, 174*, 1235–1243.

Chapter 9
Conclusion and Way Forward

Smart Technologies and Smart Cities: The Role of Public and Community Participation in Governance

Oluwayemi-Oniya Aderibigbe and Trynos Gumbo

9.1 Introduction

The term "smart city"encompasses a wide range of concepts and ideas, referring to a city that undergoes transformation by leveraging emerging technologies, such as big data, artificial intelligence, 5G, autonomous vehicles, and more (Chatigny-Vincter, 2020). According to Dameri and Cocchia (2013), the smart city represents an urban development approach aimed at improving the quality of life, enhancing services, reducing environmental impact, and promoting sustainability. Technology, particularly information and communication technology (ICT), plays a pivotal role in the smart city concept, enabling cities to efficiently manage essential functions and create added value. To effectively harness the potential of technology, it is crucial to collect data from diverse sources, including the Internet of Things (IoT), big data, and Geographic Information Systems (GIS). Marr (2015) emphasizes the significance of big data in achieving the objectives of a smart city, as it involves collecting data from various sources, such as sensors (including real-time video and "human sensors"), social media, and open data sets.

In recent times, the concept of smart cities has gained significant momentum, primarily due to their capacity to offer innovative solutions across various domains, including mobility, environment, economy, governance, quality of life, and education. These advancements are made possible by harnessing advanced Information and Communication Technologies (ICT) (refer to Fig. 9.1) (Caragliu et al., 2011). Smart cities are closely linked to the emergence of novel information technologies, such as mobile devices, the semantic web, cloud computing, and the Internet of Things (Schaffers et al., 2011). The term "smart city"was initially coined in 2005 by technology companies aiming to provide comprehensive information systems that integrate urban infrastructural operations (Harrison & Donnelly, 2011).

O.-O. Aderibigbe et al., *Emerging Technologies for Smart Cities*,
https://doi.org/10.1007/978-3-031-66943-9_9

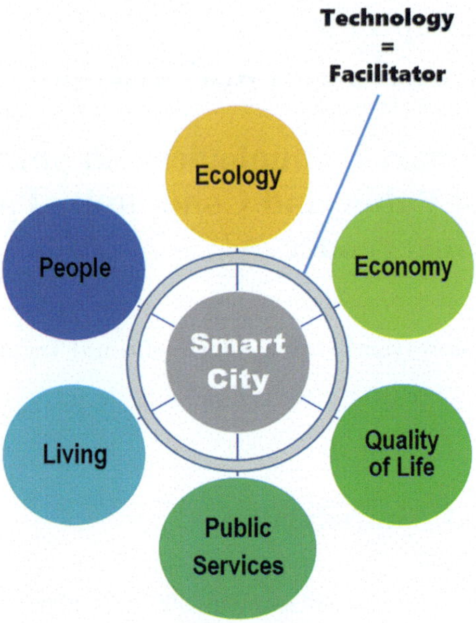

Fig. 9.1 Elements of Smart City. (Simonofski et al. 2019)

The concept of citizen participation extends beyond smart cities, but smart cities have brought attention to its significance and emphasized the importance of involving citizens in smart city initiatives and adoption. The term "smart cities"has evolved over the past 25 years and has primarily been used by Information and Communication Technology (ICT) companies such as IBM, Cisco, and Siemens. However, there is a lack of consensus on its precise meaning due to varying definitions across different sectors.

According to the Department for Business, Innovation and Skills (BIS) (2013), smart cities are a process that aims to enhance livability and resilience in urban areas. From the perspective of the UK government, a smart city should facilitate access and engagement with services in a way that meets the diverse needs of its citizens. This involves integrating infrastructure, social capital (including local skills and community institutions), and digital technologies to promote sustainable economic development and create an inclusive environment. The study identified five key features of smart cities:

- Modern digital infrastructure
- Citizen-centric service delivery
- Intelligent physical infrastructure (e.g., smart systems or the Internet of Things)
- Willingness to learn from others and experiment with new approaches and business models
- Transparency of outcomes/performance, such as city service dashboards for citizens to compare and assess performance.

However, Huber and Mayer (2015) observed that the definition and conceptual framework of smart cities lack clarity compared to concepts like low-carbon and eco cities. It remains a somewhat ambiguous concept with multiple interpretations. In their study, smart cities were conceptualized from three different perspectives:

1. **Instrumental perspective:** This perspective involves leveraging Information and Communication Technologies (ICT) to efficiently collect high-quality data from various sources in a shorter timeframe. The aim is to improve the functioning of institutions, such as municipalities. The collected data is then processed to generate meaningful insights that aid in strategic decision-making.
2. **Governance perspective:** The governance perspective emphasizes the active involvement and participation of citizens in decision-making processes. Citizens should have a significant role in shaping the future of their cities. This requires moving away from traditional top-down governance approaches and transitioning to a bottom-up approach that enables the integration of all stakeholders in decision-making. Citizen engagement is considered a crucial aspect of smart cities, along with the Internet of Things (IoT) and big data, as highlighted by Martin et al. (2019).
3. **Administrative perspective:** According to the administrative perspective, the goal of a smart city is to promote collaboration and coordination among various institutional structures and departments within a municipality. It is essential for these entities to interact and align their efforts in order to develop a unified vision for the city. This vision should be rooted in identifying and addressing the needs of the citizens.

Moreover, different scholars have presented varying definitions of a smart city. According to Vanolo (2013), the concept of a smart city originates from the ideas of "smart growth" and the "intelligent city." In this context, a smart city utilizes Information and Communication Technology (ICT) to foster its development and enhance its attractiveness. Giffinger et al. (2007) identify six key components that characterize a smart city: smart economy, smart mobility, smart governance, smart environment, smart living, and smart people. To achieve smartness, cities must invest in human and social capital while establishing both traditional (transport) and modern (ICT) communication infrastructure. This entails promoting sustainable economic growth, ensuring a high quality of life, effectively managing natural resources, and fostering participatory governance.

Mellouli et al. (2014) propose the concept of a smart community, which involves the use of ICT by local governments and cities to actively engage with citizens and leverage available data to address citizen and community concerns. The study highlights the significance of governments not only developing new services based on these technologies to enhance the quality of life for citizens but also actively involving citizens in the process of designing and implementing these services.

With the majority of the global population residing in cities, which contribute significantly to carbon emissions, there is a growing need for energy efficiency and carbon reduction. The involvement of governments and citizen participation is essential to meet challenging carbon reduction targets. Developed countries like the UK aim to achieve an 80% reduction in carbon emissions by 2050 (UNEP, 2015). While smart cities have gained attention and economic support, the role of citizens in smart city initiatives and development has often been overlooked. Many smart city projects have fallen short of their objectives due to inadequate citizen involvement and consideration of the impact on their daily lives. Therefore, it is imperative to prioritize citizen involvement and participation in order to achieve the goals of a smart city.

9.2 Theories of Participation

9.2.1 *Public Participation*

In recent years, there has been a global shift towards increased citizen engagement in policy-making, reflecting the growing interest in public participation among scholars (Bovaird, 2007). Citizen participation mechanisms have been implemented at both local and national government levels to facilitate citizens' involvement in the design, implementation, monitoring, and evaluation of public policies (Rowe & Frewer, 2005). Scholars have extensively researched this subject and developed classifications of public participation based on its purpose and level of engagement. For example, the OECD's representative typology distinguishes between information dissemination, consultation, and active participation, which involve different levels of communication and citizen involvement in decision-making processes (Gramberger, 2001).

Participation is widely acknowledged to offer numerous benefits, including advancements in democracy and legitimacy, enhancements in public policy and service quality, promotion of social inclusion and justice, as well as contributions to education and socialization processes.

9.2.2 *Co-production*

The concept of co-production, as emphasized by Hood (1995), has introduced a blend of customer-oriented and outcome-oriented thinking in policymaking and implementation. The effective utilization of information and communication technology (ICT) plays a crucial role in enhancing managerial processes and facilitating communication with citizens, thereby ensuring the successful implementation of e-Government policies. This focus on public service delivery and the involvement of citizens has given rise to the concept of New Public Governance (NPG), as highlighted by Bovaird (2007). NPG emphasizes active engagement and participation of service users in both the design and delivery of services, adopting a citizen-centric approach and promoting co-production among stakeholders.

The term "co-production," supported by scholars like Bovaird, refers to the provision of services through ongoing and collaborative relationships between professional service providers, service users, and community members. All parties involved make significant contributions to the development and progress of smart cities. While theoretical perspectives on citizen participation offer valuable insights, it is crucial to explore specific methods through which citizens can actively engage in achieving the goals of smart cities.

9.3 Participation Methods

This section offers insights into various methods of citizen participation and provides illustrative examples. According to Berntzen and Johannessen (2016), Callahan (2007), and Simonofski et al. (2017), citizen participation can be categorized into three distinct methods (refer to Fig. 9.2).

The first method is democratic participation, wherein citizens actively engage in the decision-making processes of the city, as advocated by democratic principles. This allows citizens to have a voice in shaping policies and initiatives.

The second method involves citizens acting as co-creators, contributing their ideas and proposals to enhance solutions and mitigate the risk of failure at the project's outset. By actively participating in the co-creation process, citizens become valuable collaborators in the development of smart city projects.

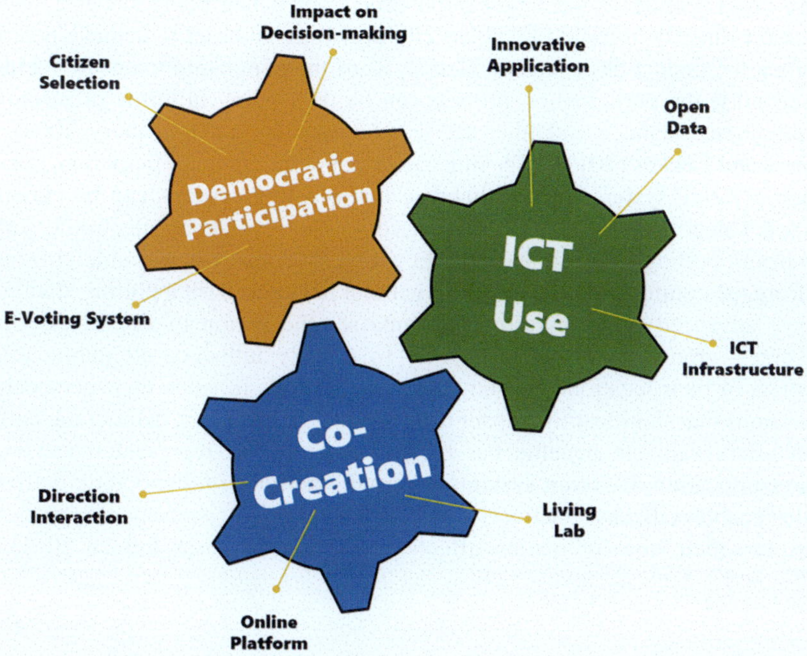

Fig. 9.2 Categories of citizen participation. (Simonofski et al., 2019)

Lastly, in the post-implementation phase, citizens can continue their participation as ICT users. By proactively utilizing the smart city infrastructure, citizens can fully immerse themselves in the technological environment, facilitating their ongoing involvement and making participation more accessible.

These three methods of citizen participation offer different avenues for individuals to engage with and contribute to the development and progress of smart cities.

9.3.1 Democratic Participation

Irvin and Stansbury (2004) posit that citizen participation in the decision-making process can be regarded as a manifestation of democratic engagement. This approach presents numerous benefits, including the empowerment of citizens to acquire knowledge on intricate technical matters and develop expertise in areas pertinent to the public interest. Additionally, public officials stand to gain valuable insights from citizens, thereby obtaining information about the advantages and disadvantages associated with various policies. Notably, democratic participation proves to be a cost-effective approach as it diminishes the likelihood of litigation or investments that are wasteful or not aligned with public preferences.

Nevertheless, democratic participation does have its limitations. Firstly, it is essential to ensure that the group of citizens engaged in the process represents a diverse cross-section of the larger population, rather than being skewed towards those most directly impacted by smart city decisions. In order to achieve this, it is important to conduct basic statistical analysis of the population to ensure inclusivity. Secondly, the participation process can be demanding in terms of resources, including time, finances, and other necessary means (Irvin & Stansbury, 2004).

Zissis and Lekkas (2011) offer illustrations of democratic participation, notably through e-voting systems. They elucidate that e-voting systems can be classified into two categories: physically supervised systems, such as electronic voting machines in polling stations, and remote voting systems, which enable citizens to vote from the comfort of their homes or without having to visit a polling station. At its most advanced stage, e-voting encompasses the notion of Electronic Direct Democracy, granting citizens the ability to directly influence all public affairs regardless of their location. This empowers individuals to voice their perspectives on decisions that significantly impact their welfare. In summary, democratic participation offers valuable benefits but requires attention to representativeness and resource allocation. E-voting systems represent one form of democratic participation that enables citizens to actively participate in public decision-making processes and express their views on matters affecting their lives (Zissis & Lekkas 2011).

9.3.2 Citizens as Co-creators

The traditional approach to urban innovation involved centralized decision-making by urban planners based on their own ideas and perspectives. However, in the context of smart cities, a new model has emerged that harnesses the input and ideas of citizens (Schaffers et al., 2011). Citizens are no longer passive consumers but are recognized as crucial stakeholders who can generate valuable ideas and address social needs. Various methods are employed to enable citizens as co-creators, including focus groups, expert interviews, town hall meetings, usability testing, real-time comments, and the development of service quality standards (Johannessen, 2010).

In addition to these approaches, gathering citizen input for e-government services can be achieved through requirements engineering, which aims to accurately capture the goals, needs, and expectations of citizens. Two methods for enhancing citizen participation are centralized platforms and social media analysis (Berntzen & Johannessen, 2016). While developing and maintaining centralized platforms can be costly and challenging, social media offers an opportunity to engage with a larger number of citizens in various contexts. Crowdsourcing platforms, collaboration tools, social networking platforms, and online surveys are examples of how social media can facilitate citizen engagement (Criado et al., 2013).

Canteneur (2015) highlights the existence of other platforms that collect citizens' opinions and experiences on public matters. One such platform is "Civiocracy," an online platform that involves citizens, companies, NGOs, and political authorities in discussions on a wide range of issues, from poverty to specific community needs. Civiocracy stimulates citizens' engagement in public life and encourages them to provide their unique insights on various issues.

In summary, the shift towards citizen co-creation in smart cities recognizes the value of citizen input and utilizes various methods such as focus groups, usability testing, and social media engagement to involve citizens in decision-making processes and gather their perspectives and ideas. Online platforms like Civiocracy further promote citizen engagement and enable them to contribute their insights on public matters (Canteneur, 2015).

9.3.3 Citizens as ICT Users

Simply incorporating information and communication technology (ICT) into smart cities is insufficient, and an overemphasis on ICT has been recognized as a significant limitation in certain smart city initiatives (Merli & Bonollo, 2014). However, the integration of ICT within a city can unlock a plethora of new opportunities and reshape its landscape. Friedewald and Raabe (2011) propose that technological progress facilitates the establishment of a "ubiquitous computing" infrastructure, characterized by the integration of wireless, interconnected microprocessors into everyday objects, enabling them to sense and alter the environment. The crucial aspect lies in harnessing these technological advancements to benefit the citizens, although these developments might still seem abstract to the general public, who are primarily concerned with practical solutions (Schaffers et al., 2011). To address this, new citizen-oriented applications can be developed and aligned with the existing infrastructure. These innovative applications span from Augmented Reality systems to Citizen Science platforms, Public Displays, and any other application that enables citizens to feel surrounded by and supported by technology, thus motivating their engagement (Gutierrez et al., 2013; Du et al., 2017; Khan & Kiani, 2012). Open data is an example of citizens functioning as ICT user

- **Open Data:** Open data refers to publicly generated data that is shared without restrictions (Janssen et al., 2012). It promotes the concept of governments operating as transparent systems, actively engaging with the environment and embracing diverse perspectives and feedback. Open data covers a wide range of domains, including traffic, weather, public sector budgeting, and tourist information. However, the mere publication of open data does not automatically guarantee citizen participation, as it necessitates significant transformations within the public sector and the development of citizen skills to effectively utilize this data. Nevertheless, proactive citizens can establish open-source platforms to harness open data, fostering collaboration among citizens and empowering them to address issues at various levels, from local neighborhoods to entire nations.

9.4 Role of ICT in Citizen Participation

ICT has emerged as a valuable tool for promoting and enhancing public participation (Ishikawa, 2002). For example, Marres (2012) emphasizes that ICT can lower the barriers to participation by allowing citizens to engage through their mobile devices at their convenience and from any location. The various modes of expression and communication facilitated by ICT also cultivate citizens' interest and legitimacy in participating in public affairs. In addition to broadening public participation, ICT has the potential to enhance the quality of citizens' input, which is no longer restricted to verbal forms alone. Collaborative tools, such as citizen sensing and interactive applications, have been identified by Gutierrez et al. (2013) as means to enrich democratic discussions.

Furthermore, e-participation, which leverages ICT for participatory processes, enhances citizens' ability to engage in the political process and make their opinions known during decision-making (Gramberger, 2001). ICT tools not only enable citizens to express their opinions on issues but also provide them with access to information on public policies, empowering them to co-create these policies. The exchange between governments and citizens includes the sharing of information, seeking consultation, and involving citizens in political decision-making processes. ICT plays a vital role in enabling these interactions and has the capacity to rebuild trust in government (Hague & Loader, 1999). In the digital realm, citizens can engage with public officials in a less formal manner, fostering horizontal and egalitarian interactions instead of vertical and bureaucratic ones (Macintosh et al., 2009).

9.4.1 The Requirements of ICT Participation

Berntzen and Karamagioli (2010) have developed a model (see Fig. 9.3) that outlines the prerequisites for citizen participation through ICT. One crucial requirement is the availability of relevant infrastructure, which can be wired or wireless, and serves as the foundation for establishing connections between citizens and participatory platforms. Additionally, citizens need access to technology, whether through personal devices such as computers, tablets, or smartphones, or through publicly accessible devices like information kiosks or library computers.

Another essential prerequisite for citizen participation is training and awareness. Often, citizens may not be fully aware of the importance of their participation and involvement in the design of a smart city, thus emphasizing the need for raising

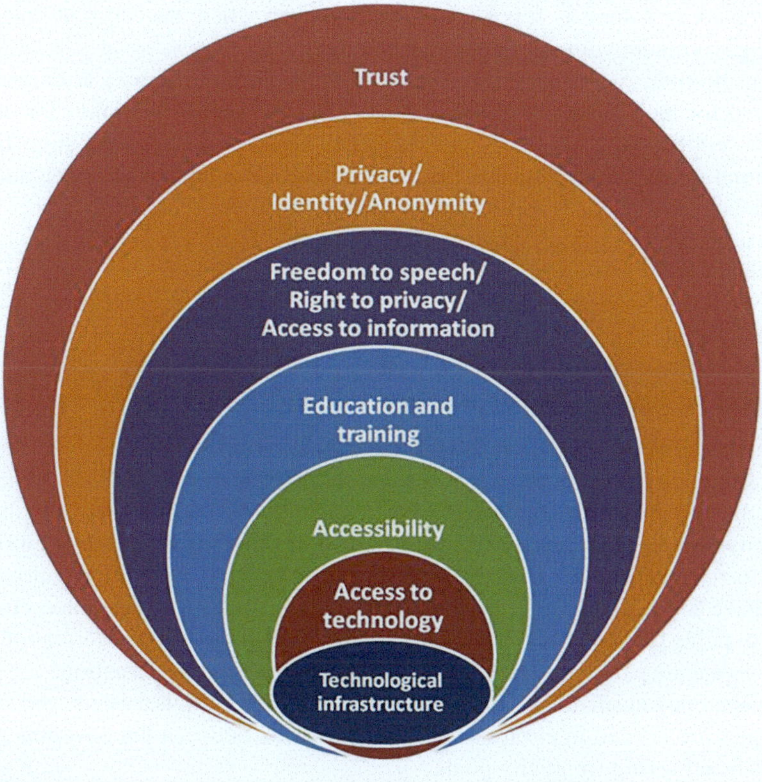

Fig. 9.3 Process of Citizen Participation via ICT use. (Berntzen & Johannessen, 2016)

awareness. Consequently, training and education become necessary to enable citizens to utilize technology efficiently and meaningfully. Legal mechanisms are also required to protect privacy, freedom of expression, and other rights. Technical mechanisms should be in place to ensure user authentication when necessary and to provide anonymity in other contexts. These layers of infrastructure, technology access, training and awareness, and legal and technical mechanisms are all indispensable for fostering trust in and achieving widespread adoption of smart cities.

9.5 Political Participation and Smart Cities Development

According to the Organisation for Economic Cooperation and Development (OECD, 2001), maintaining interaction between government and citizens at all levels is crucial for the smooth development of smart cities. To understand this interaction, a model focusing on the levels of engagement between government and citizens has been developed, encompassing three distinct levels:

(a) Information Only: This level involves one-way communication where the government informs citizens about political and managerial issues without seeking feedback or response. It is a top-down approach where decisions are made by the government without proper consultation.
(b) Consultation: At this level, there is two-way communication, and citizens are asked for their input on specific issues. The city selects the topics for discussion, collects and processes the feedback. This form of participation fosters mutual relationships between the government, relevant stakeholders, and citizens, as everyone is involved in the decision-making process.
(c) Participation: This level entails citizens entering a partnership with the city and actively participating in political decision-making. Citizens have the opportunity to raise issues they want to discuss, and the city listens and considers their input. While the OECD model discusses interactions in general, without specific emphasis on digital interaction, incorporating digital tools can enhance the interaction between the city and the government.

In order for citizen engagement in political decision-making to effectively contribute to smart city initiatives, it is crucial for citizens to possess comprehensive knowledge about their municipality's past, present, and future developments. The smart city should prioritize the dissemination of up-to-date and pertinent information to foster transparency. Lidberg (2009) emphasizes that democratic nations generally have legal frameworks and mechanisms in position to guarantee citizens' access to public documents and records. Smart cities can leverage information technology to enhance the transparency and availability of such information.

Transparency in smart city governance encompasses various aspects, and several techniques are needed to ensure a smooth interaction between the government and citizens in achieving smart city goals:

- Document transparency involves providing access to papers used as a basis for decision-making. Municipalities are required to maintain records of incoming and outgoing mail, including emails, which are generally public unless they contain privacy-protected information. Documents can be made available on the municipal or city website for citizens to read and download.
- Meeting transparency focuses on providing access to meetings where decisions are made. Detailed information about the venue, time, agenda, etc., should be provided. Information from these meetings can be disseminated through email newsletters and municipal or city websites. Webcasting meetings can also enable more citizens to follow the proceedings, and publishing minutes of political meetings online ensures their accessibility.
- Process transparency aims to help citizens understand political decision-making and the various steps involved. Information technology can play a significant role in visualizing decision-making processes, such as using flowcharts and timelines to demonstrate the steps and progress.
- Benchmarking transparency involves comparing the performance of the municipality or city with others. Statistical information can be submitted to databases like KOSTRA in Norway, facilitating benchmarking using relevant indicators. Other benchmarking websites, like BedreKommune.no by the Norwegian Association of Local and Regional Authorities (KS), provide additional indicators for comparison.
- Decision-maker transparency is crucial for building trust. Information about ownership or other interests should be disclosed to the public. In Norway, a register maintained by KS allows public access to such information.
- Disclosure transparency ensures citizens have the right to ask questions about information not included in documents or proceedings. Mechanisms such as question sessions with the mayor, where citizens can ask questions that are answered in local council meetings, or the mayor being available at public places like the library, facilitate discussions. ICT can provide an efficient channel for such interactions.

By implementing these transparency techniques, smart cities can enhance citizen participation, foster trust, and promote effective collaboration between the government and citizens in the realization of smart.

9.5.1 Tools of Participation

OECD type	Tool	Features
Information	Online email	According to Berntzen et al. (2009), most communities are required to keep records of incoming/outgoing mail. This will allow citizens to access and examine both past and present records of policies being made on their behalf by the government and relevant stakeholders. In this way, they have access to vital documents which makes it easier for citizens to be involved in decision making.
Information	Online document repository	Most municipalities provide access to all documents used for political decision making (document repository) including meeting agendas and minutes from the meetings. Online access makes it easier for citizens to view all relevant documents related to the decision-making process (Berntzen et al., 2009). For example, all Norwegian municipalities have their own web sites providing information about municipal services and organizations.
Information	Webcasts	Another tool of participation is Webcasts, and through this method, citizens may follow the proceedings of the meetings of political bodies. The proceedings may be recorded and made available for future viewing (Berntzen et al., 2009) Webcasting of local council meetings is also a popular service.
Consultation	Consultations	Consultations can be seen as a structured way of getting feedback on policy issues. Electronic consultations are easy to implement, and citizens can be asked to provide their responses and opinions via e-mail. Another way of consulting with the citizens is through the use of modified discussion forum platforms to facilitate sharing and dialog. Citizens should receive feedback on their input, if not interest in participating will decrease. Several communities have used this method as a way of providing consultations on political matters, and citizens are allowed to comment on specific policy proposals and plans. An example is "Digital Planning Dialog" (Berntzen & Trollvik, 2007). This allows citizens to examine both maps and documents related to spatial planning, and also make comments about the plans.
Consultation	Polling	Polling is also a form of consultation where citizens are asked about specific issues. In a study by Goidel (2011), Online polling applications normally uses scales or checkboxes to collect opinions from the citizens.
Consultation	Blogs	The city may also use a blog to ask for input on matters of public interest. A blog is a collection of entries containing text/photos/videos with the latest entry shown first. The blog normally allows readers (citizens) to comment.
Participation	Discussion forum	A discussion forum is a platform that typically allows its users to start threads on some topic (Berntzen, 2004). Other users may add their comments.

OECD type	Tool	Features
Participation	Social network platforms (e.g., Facebook)	Many municipalities are using social media platforms for information, mobilization and dialog. Facebook and Instagram have been recorded as one of the most widely used. The pages are mostly used for information, but some municipalities encourage questions and provide answers. Some municipalities are using twitter and other social media platforms to generate information from citizens as a way of citizen participation and involvement in decision making.
Participation	Participatory budgeting	Participatory budgeting has become widespread across the globe and is a situation where the citizens vote on the use of (a portion) of the total budget for a city. In this case, the results are a consequence of the participation (Sintomer et al., 2016). Participatory budgeting is a powerful mechanism to make participation work. The incentive to participate is high, since the citizens will see direct results from taking part in the decision making.

Source: Berntzen and Johannessen (2016)

9.6 Benefits of Citizen Participation in Smart Cities

Public participation in governance has been a prominent topic in public administration research for over 50 years, influenced by social movements of the 1960s and 1970s (Dean, 2017; Lember et al., 2022). Ostrom (1996) highlights that public participation is increasingly recognized as a means to govern commons and enhance the effectiveness of public policy making and service delivery. Engaging citizens and service users allows governments to tap into their valuable knowledge and external resources regarding policy and service aspects. Citizen participation provides a platform for comprehensive governance that strengthens democracy and gives citizens a voice on issues that affect their daily lives and well-being (Hao et al., 2022).

In the context of smart cities, citizen engagement plays a crucial role in promoting effective decision-making and maximizing the impact of investments. By involving citizens in the decision-making process, encouraging them to contribute ideas, express concerns, and influence investment decisions, the overall outcome can be significantly enhanced (Gilbreath Holdar & Zakharchenko, 2002). Citizen participation refers to the collaborative process through which citizens organize themselves and work together through non-governmental community organizations to influence decision-making processes at the grassroots level (Gilbreath Holdar & Zakharchenko, 2002). Citizens can actively participate in local governance by expressing their needs and actively engaging in local project planning and budgeting (Day, 1997; Ebdon, 2002; Tanaka, 2007). There are various avenues for public participation, including public hearings, meetings, campaigns, civil associations, citizen advisory committees, and petitions (Gilbreath Holdar & Zakharchenko, 2002; Callahan, 2007).

Citizen participation in smart cities offers several benefits. Firstly, it educates citizens and promotes a sense of citizenship, contributing to building a progressive society and promoting sustainable development through the adoption of smart city initiatives (Team, 2015). Secondly, citizen participation provides ideas on public issues, supports government decision-making, mitigates conflicts and delays, and fosters cooperation between the government and the public (Marzuki, 2015; Hao et al., 2022). It helps analyze policies and their impact, gain public interest early in the decision-making process, and enables citizens to understand the decision-making process (Marzuki, 2015; Hao et al., 2022).

Moreover, citizen participation enhances the efficiency and effectiveness of government policies and services. By involving citizens at every stage of decision-making and initiatives, governments can improve policy implementation, decision-making quality, and overall governance (Sidor, 2012). Citizen participation also creates a platform for developing creative and innovative solutions to environmental problems from diverse perspectives, leading to improved environmental protection (Li et al., 2018). Furthermore, citizen participation builds trust in elected leaders, empowers citizens, and contributes crucial information for policymaking (Fung, 2015). It enables authorities to make informed and inclusive choices, programs, and rules, mitigating conflicts that may arise from missed consultations and non-inclusive policymaking (Fung, 2015).

According to Bull (2016), citizen participation methods have the potential to foster a robust and transparent democratic process, while also benefiting municipalities by promoting innovative and sustainable solutions, strengthening local action, and garnering support for the implementation of new ideas. In the realm of investments, participatory budgeting can facilitate inclusive and creative decision-making throughout the entire process, from selecting investment priorities to providing feedback on the outcomes achieved. Overall, Grimshaw and Lever (2008) highlight several ways in which public participation contributes to the realization of smart cities:

- Enhanced self-esteem of participants.
- Increased knowledge and awareness of civil and political aspects of life.
- Heightened awareness and understanding of individual self-interests.
- Opportunity for expression of personal social identity, such as socialist, conservative, feminist, internationalist, etc.
- Strengthened social citizenship.
- Improved managerial efficiency by incorporating a wider range of perspectives pertaining to societal well-being, thus serving as a valuable reality check.

In summary, citizen participation allows governments to arrive at better solutions for their problems, promotes accountability to elected leaders, and takes into account the relevant needs of society. It upholds democratic values of efficiency, legitimacy, and social justice. The significant changes brought about by advanced technology have heightened the demand for participatory management systems that are considered competent, while traditional hierarchical systems of governance, primarily characterized by government anarchy, have faced criticism. In today's society, citizens expect governments to be designed for and by the people. A government that does not adopt a democratic form is deemed illegitimate, as power in a democratic society emanates from the people. Citizen participation serves to strengthen the relationship between citizens and government.

9.7 Challenges of Citizen Participation in Smart City Initiatives

The implementation of public participation processes is crucial for achieving sustainable goals and meeting public needs. However, numerous factors can impede the successful implementation of public participation in smart city initiatives. Peng (2020) identifies several key factors:

- **Cost:** Including citizens in decision-making processes can be financially and resource-intensive for governments. It requires adequate financial and human resources, as well as significant time investment for meaningful engagement. Arranging forums, workshops, and public meetings demands sufficient funds, preparation time, and staffing. The transaction cost of participation projects is often higher compared to the benefits, posing a challenge to conducting low-cost participation projects.
- **Inadequate skilled facilitators:** To engage citizens effectively, skilled facilitators are essential. They can educate and train citizens on the importance of their input and contributions to smart city projects. Additionally, facilitators may need to provide education on the use of relevant technologies for achieving smart mobility. However, some administrators organizing public engagement lack the necessary skills. Skilled practitioners are crucial for facilitating the process, maintaining positive citizen relationships, and generating better outcomes. Training programs are available, but they can be expensive and may not consider the specific context.
- **Low efficiency and inadequate time on the part of citizens:** Citizen participation often requires a significant time commitment, which can be challenging when balancing work, household obligations, and other daily tasks. Traditional public engagement approaches can be time-consuming with less satisfactory results, discouraging participation. Increasing the efficiency of participation approaches without compromising outcomes is an urgent need.
- **Lack of interest and dedication on the part of citizens:** Public interest in public matters may be low, affecting participation levels. Motivating citizens, stimulating their interests, and obtaining meaningful inputs are challenges that practitioners must address. Encouraging participation through various monetary or non-monetary rewards and providing opportunities related to citizens' daily lives can help increase participation rates.
- **Lack of expertise:** Citizens often lack specialized expertise, especially regarding complex technologies. However, understanding the technical details of

advanced technologies is not essential for grasping their crucial social dimensions. Educating citizens about the usage and implications of these technologies can help them develop a better understanding of key issues.

- **Inadequate awareness:** Many citizens may be unfamiliar with the issues that need to be addressed in smart city adoption and decision making. Raising public awareness can help address the problem of inadequate knowledge, but it may not solve the challenge of limited time for everyone to become knowledgeable about all issues. Decision-makers need to create sufficient opportunities to inform and involve citizens in the issues affecting their communities and well-being.

In summary, citizen involvement is essential for driving decisions in smart city initiatives, considering that citizens are the end-users of the products and services. However, both internal and external factors influence decision-making behavior. Smart city initiatives have the potential to overcome the limitations of traditional urban development by leveraging digital technologies and fostering collaboration among stakeholders, improving citizen involvement, offering new services, and generating innovative ideas for city operations.

9.8 Conclusion and Way Forward

Emerging technology are becoming favorable in the transportation system of any community in achieving a smart city, however, there are challenges mitigating effective utilization and adoption of these technologies especially in major cities of the global south. Technology plays a pivotal role in optimizing value chains, minimizing resource consumption and waste, and facilitating the dissemination of climate and real-time weather data. It also contributes significantly to forecasting early warning systems and supporting resilience and climate adaptation efforts. Digital solutions with sustainability benefits, such as smart agriculture, smart building, smart energy, smart manufacturing, and smart mobility, enable the reduction of greenhouse gas emissions (GHG) and promote the adoption of renewable energy sources, thereby driving market transformation. Furthermore, through the implementation of forecasting and early warning systems, Information and Communication Technology (ICT) effectively monitors climate change and strengthens resilience by aiding in the mitigation of its adverse effects.

Nevertheless, there are barriers that hinder the widespread adoption of these technologies, notably the absence of a systematic approach to bolster Intelligent Transportation Systems (ITS) in India. Efficient utilization of ITS necessitates the integration of diverse technologies and applications across all road types in the country. Currently, ITS applications in India are underutilized, indicating a pressing need for further efforts to harness their potential. To address this, greater emphasis should be placed on enhancing the utilization of ITS applications, accompanied by a concerted focus on diverse services and applications. Consequently, there is a compelling need for coordinated efforts among various departments to effectively implement technology in transport planning initiatives.

Additionally, fostering citizen participation and engagement is crucial for effectively leveraging technology to achieve sustainable development goals and realize the vision of smart cities. Strengthening existing infrastructure to enhance safety, inclusivity, and resilience is also imperative. ICT platforms facilitate participatory urban planning initiatives, enabling citizens to voice their opinions and aiding governments and relevant stakeholders in better tailoring city planning to meet the specific needs of users. Sustainable Development Goals 1 and 11 (SDG1 and 11), targeting poverty eradication and comprehensive city management, are of paramount importance in this endeavor. The adoption of innovative technology is perceived as pivotal in unleashing significant human potential, particularly in many Global South countries, thereby enhancing their international competitiveness and positively impacting the global economy's development.

References

Berntzen, L. (2004). Content analysis of a local government initiated discussion forum. In R. Traunmüller (Ed.), *Electronic government—Proceedings of the third international conference, EGOV 2004, Zaragoza, Spain, August 30–September 3, 2004* (pp. 300–307). Springer.

Berntzen, L., & Johannessen, M. R. (2016). The role of citizen participation in municipal smart city projects: Lessons learned from Norway. In *Smarter is the new urban agenda* (pp. 299–314). Springer International Publishing.

Berntzen, L., & Karamagioli, E. (2010) *Regulatory measures to support eDemocracy*. Proceedings Fourth International Conference on the Digital Society (ICDS) (pp. 311–316). https://doi.org/10.1109/ICDS.2010.74, IEEE Press.

Berntzen, L., & Trollvik, J. A. (2007). Digital planning dialog. In *Proceedings of the third international conference on e-government, Montreal, Canada* (pp. 61–68). Academic Conferences Limited.

Berntzen, L., Nietzio, A., Goodwin Olsen, M., AbdelGawad, A., & Snaprud, M. (2009) *State-of-the-art review: Transparency indicators*. Deliverable number D2.1.2. eGovMon project. http://archive.tingtun.no/eGovMon/d2.1.2-final-v1.0.pdf

BIS (Department for Business, Innovation and Skills). (2013). *Smart cities: Background paper*. BIS. https://www.gov.uk/government/uploads/system/uploads/attachment_data/file/246019/bis-13–1209-smart-cities-background-paper-digital.pdf. Accessed 22 Oct 2020.

Bovaird, T. (2007). Beyond engagement & participation: User & community co-production of public services. *Public Administration Review, 67*(5), 846–860.

Bull, R. (2016). Smart citizens for smart cities: Participating in the future. *Proceedings of the Institute of Civil Engineers-Energy, 169*(3), 93–101. https://doi.org/10.1680/jener.15.00030

Callahan, K. (2007). Citizen participation: Models and methods. *International Journal of Public Administration, 30*(11), 1179–1196. https://doi.org/10.1080/01900690701225366

Canteneur, P. (2015). *Civiocracy, une plateforme intelligente pour débattre de sujets civiques*. Retrieved November 7, 2020, from http://www.atelier.net/trends/articles/civocracy-une-plateforme-intelligente-debattre-de-sujets-civiques_433992

Caragliu, A., Del Bo, C., & Nijkamp, P. (2011). Smart cities in Europe. *Journal of Urban Technology, 18*(2), 65–82.

Chatigny-Vincter, A. (2020). *Public participation and inclusion in smart city projects in Montreal*. A thesis submitted for B.A. Honours Degree in Geography, McGill University, Montreal.

Criado, J. I., Sandoval-Almazan, R., & Gil-Garcia, J. R. (2013). Government innovation through social media. *Government Information Quarterly, 30*(4), 319–326.

Dameri, R. P., & Cocchia, A. (2013). *Smart city and the digital city: Twenty years of terminology evolution*. Proceedings Itais 2013. http://www.cersi.it/itais2013/pdf/119.pdf

Day, D. (1997). Citizen participation in the planning process: An essentially contested concept? *Journal of Planning Literature, 11*(3), 421–434. https://doi.org/10.1177/088541229701100309

Dean, R. (2017). Beyond radicalism and resignation: The competing logics for public participation in policy decisions. *Policy and Politics, 45*(2), 213–230.

Du, G., Degbelo, A., & Kray, C. (2017). *Public displays for public participation in urban settings: A survey*. Proceedings of the 6th ACM international symposium on pervasive displays, 17.

Ebdon, C. (2002). Beyond the public hearing: Citizen participation in the local government budget process. *Journal of Public Budgeting, Accounting and Financial Management, 14*(2), 273–294. https://doi.org/10.1108/JPBAFM-14-02-2002-B006

Friedewald, M., & Raabe, O. (2011). Ubiquitous computing: An overview of technology impacts. *Telematics and Informatics, 28*(2), 55–65.

Fung, A. (2015). Putting the public back into governance: The challenges of citizen participation and its future. *Public Administration Review, 75*(4), 513–522.

Giffinger, F. Fertner, C., Kramar, H., Kalasek, R. Pichler-Milanovic, N., & Meijers, E. (2007). Smart cities: Ranking of European medium-sized cities. *Final report for smart-cities.eu*. URL: www.smart-cities.eu/download/smart_cities_final_report.pdf

Gilbreath Holdar, G., & Zakharchenko, O. (2002). *Citizen participation handbook*. iMedia Ltd.

Goidel, K. (Ed.). (2011). *Political polling in the digital age—The challenge of measuring and understanding public opinion*. Louisiana State University Press.

Gramberger, M.R. (2001). *Citizens as partners: OECD handbook on information, consultation and public participation in policy-making*, OECD.

Grimshaw, L. & Lever, J. (2008). *Citizens' participation in policy-making: Working paper*. Accessed on January 10 2022 from https://pure.hud.ac.uk/ws/portalfiles/portal/14275910/Working_paper_WP14_draft_080909_1.pdf

Gutierrez, V., Galache, J. A., Sanchez, L., Munoz, L., Hernandez-Munoz, J. M., Fernandes, J., & Presser, M. (2013). SmartSantander: Internet of things research and innovation through citizen participation. *Lecture notes in computer science* (including subseries lecture notes in artificial intelligence and lecture notes in bioinformatics), 7858 LNCS, pp. 173–186.

Hao, C., Nyaranga, M. S., & Hongo, D. O. (2022). *Enhancing public participation in governance for sustainable development: Evidence from Bungoma County*. Sage Open Access. https://doi.org/10.1177/21582440221088855

Harrison, C., & Donnelly, I. (2011). A theory of smart cities. In *Proceedings of the 55th annual meeting of the ISSS*.

Hood, C. (1995). The 'new public management' in the 1980s: Variations on a theme. *Accounting Organization and Society, 20*(2/3), 93–109.

Huber A & Mayer I (2015). *Is this a smart city? Narratives of city smartness and their critical assessment*. Proceedings of the 2015 ECEE summer study conference, Buffalo, NY, USA, pp. 817–823.

Irvin, R. A., & Stansbury, J. (2004). Citizen participation in decision making: Is it worth the effort? *Public Administration Review, 64*(1), 55–65.

Ishikawa, Y. (2002). Calls for deliberative democracy in Japan. *Rhetoric & Public Affairs, 5*(2), 331–345.

Janssen, M., Charalabidis, Y., & Zuiderwijk, A. (2012). Benefits, adoption barriers and myths of open data and open government. *Information Systems Management, 29*(4), 258–268.

Johannessen, M. (2010). *Genres of participation in social networking systems: A study of the 2009 Norwegian parliamentary election*. In Lecture notes in computer science (including subseries lecture notes in artificial intelligence and lecture notes in bioinformatics) (Vol. 6229 LNCS, pp. 104–114).

Khan, Z., & Kiani, S. L. (2012). *A cloud-based architecture for citizen services in smart cities*. Proceedings – 2012 IEEE/ACM 5th international conference on utility and cloud computing, UCC 2012, pp. 315–320.

Lember, V., Randma-Liiv, T., & Vooglaid, K. M. (2022). Engaging citizens in policy making: The potential and challenges of e-participation. In V. Lember & T. Randma-Liiv (Eds.), *Political Science and Public Policy*. https://doi.org/10.4337/9781800374362.00006

Li, L., Xia, X. H., Chen, B., & Sun, L. (2018). Public participation in achieving sustainable development goals in China: Evidence from the practice of air pollution control. *Journal of Cleaner Production, 201*, 499–506. https://doi.org/10.1016/j.jclepro.2018.08.046

Lidberg, J. (2009). *The international freedom of information index—The promise and practice of FOI laws*. VDM Vorlag Dr. Müller.

Hague B.M, & Loader, B. (1999). Digital democracy: Discourse and decision making in the information age, Routledge.

Macintosh, A., Coleman, S., & Schneeberger, A. (2009). E-participation: The research gaps. *Electronic Participation, 5694*, 1–11.

Marr, B. (2015). *Big data—Using smart big data to make better decisions and improve performance*. Wiley.

Marres, N. (2012). *Material participation: Technology, the environment and everyday publics*. Palgrave Macmillan.

Martin, C., Evans, J., Karvonen, A., et al. (2019). Smart-sustainability: A new urban fix? *Sustainable Cities and Society, 45*, 640–648.

Marzuki, A. (2015). Challenges of public participation and the decision-making process. *Sociology and Space, 53*(1), 21–39.

Mellouli, S., Luna-Reyes, L. F., & Zhang, J. (2014). Smart government, citizen participation and open data. *Information Polity, 19*(1–2), 1–4.

Merli, M., & Bonollo, E. (2014). Performance measurement in the smart cities. In *Smart City : How to create public and economic value with high technology in urban space* (pp. 139–155). Springer.

Organisation for Economic Cooperation and Development (OECD). (2001). *Engaging citizens in policy-making, in OECD public management policy brief*. OECD.

Ostrom, E. (1996). Crossing the great divide: Coproduction, synergy, and development. *World Development, 24*(6), 1073–1087.

Peng, Y. (2020). *5 principal challenges of public participation*. Accessed on February 5, 2023 from https://wesolve.app/5-principle-challenges-of-public-participation/

Rowe, G., & Frewer, L. J. (2005). A typology of public engagement mechanisms. *Science, Technology & Human Values, 30*(2), 251–290.

Schaffers, H., Komninos, N., Pallot, M., Trousse, B., Nilsson, M., & Oliveira, A. (2011). *Smart cities and the future internet: Towards cooperation frameworks for open innovation*. Lecture Notes in Computer Science (Including Subseries Lecture Notes in Artificial Intelligence and Lecture Notes in Bioinformatics), 6656, pp. 431–446.

Sidor, M. (2012). The process of enhancing citizens' participation in local government in Poland. *Socialiniai Tyrimai, 3*(28), 87–97.

Simonofski, A., Serral Asensio, E., Desmedt, J., & Snoeck, M. (2017). Citizen Participation in Smart Cities: Evaluation Framework Proposal. 2017 IEEE 19th Conference on Business Informatics (CBI), 227–236.

Simonofski, A., Asensio, E. S., & Wautelet, Y. (2019). Citizen participation in the design of smart cities: Methods and management framework. Retrieved on November, 2021 from https://www.researchgate.net/publication/333939575_Citizen_participation_in_the_design_of_smart_cities?enrichId=rgreq-737238f480fd462381ade2a359d3ff94-XXX&enrichSource=Y292ZXJQYWdlOzMzMzkzOTU3NTtBUzo5OTkwMjQ0ODQxNjM1ODZAMTYxNTE5NzQwNjMzNA%3D%3D&el=1_x_2&_esc=publicationCoverPdf

Sintomer, Y., Röcke, A., & Herzberg, C (2016). *Participatory Budgeting in Europe: Democracy and public governance*. Routledge: New York, USA.

Tanaka, S. (2007). Engaging the public in national budgeting: A non-governmental perspective. *OECD Journal of Budgeting, 7*(2), 139–177. https://doi.org/10.1787/budget-v7-art12-en

Team, I. P. A. (2015). Public participation and citizen engagement: Effective advising state-building contexts: How, Geneva, an initiative inter peace. IPA.

UNEP (2015). The Emissions Gap Report 2015. United Nations Environment Programme (UNEP), Nairobi. Asssessed on December, 2020 from http://www.unep.org/emissionsgapreport2015/

Vanolo, A. (2013). Smart mentality: The Smart City as Disciplinary Strategy. *Urban Studies 51*(5):883–898. https://doi.org/10.1177/0042098013494427.

Zissis, D., & Lekkas, D. (2011). Securing e-Government and e-Voting with an open cloud computing architecture. *Government Information Quarterly, 28*(2), 239–251.

Acronyms

ALPR	Automatic License Plate Recognition
AT	Active Travel
CNG	Compressed Natural Gas
DSRC	Dedicated Short Range Communication
ETC	Electronic Toll Collection
FERMA	Federal Roads Maintenance Agency
FHWA	Federal Highway Administration
FRSC	Federal Road Safety Commission
GHG	Green House Gas
GIS	Geographic Information System
HAR	Highway Advisory Radio
ICT	Information and Communication Technology
IoT	Internet of Things
ITS	Intelligent Transportation Systems
LRT	Light Rail Transit
LTA	Land Transport Authority
MCDA	Multi-criteria Decision Analysis
MRT	Mass Rapid Transit
NHTS	National Household Travel Survey
NMT	Non-motorised Transport
OBU	On-Board Unit
OECD	Organisation for Economic Co-operation and Development
RTS	Rapid Transit System
SeP	Symphony for e-Payment
TCIP	Tin Can Island Port
UAV	Unmanned Aerial Vehicle
UN	United Nations
VMS	Variable Message Signs
VPS	Vehicle Positioning System

O.-O. Aderibigbe et al., *Emerging Technologies for Smart Cities*,
https://doi.org/10.1007/978-3-031-66943-9

V2V	Vehicle-to-Vehicle technology
V2I	Vehicle-to-Infrastructure technology
WHO	World Health Organization

Index

O.-O. Aderibigbe et al., *Emerging Technologies for Smart Cities*,
https://doi.org/10.1007/978-3-031-66943-9